PHILOSOPHY OF SCIENCE

Other interview books from Automatic Press ◆ $\frac{V}{I}$P

Formal Philosophy
edited by Vincent F. Hendricks & John Symons
November 2005

Masses of Formal Philosophy
edited by Vincent F. Hendricks & John Symons
October 2006

Political Questions: 5 Questions for Political Philosophers
edited by Morten Ebbe Juul Nielsen
December 2006

Philosophy of Technology: 5 Questions
edited by Jan-Kyrre Berg Olsen & Evan Selinger
February 2007

Game Theory: 5 Questions
edited by Vincent F. Hendricks & Pelle Guldborg Hansen
April 2007

Philosophy of Mathematics: 5 Questions
edited by Vincent F. Hendricks & Hannes Leitgeb
January 2008

Philosophy of Computing and Information: 5 Questions
edited by Luciano Floridi
Sepetmber 2008

Epistemology: 5 Questions
edited by Vincent F. Hendricks & Duncan Pritchard
September 2008

Mind and Consciousness: 5 Questions
edited by Patrick Grim
January 2009

Epistemic Logic: 5 Questions
edited by Vincent F. Hendricks and Olivier Roy
August 2010

Philosophy of Physics: 5+1 Questions
edited by Juan Ferret & John Symons.
November 2010

See all published and forthcoming books in the 5 Questions series at
www.vince-inc.com/automatic.html

PHILOSOPHY OF SCIENCE
5 QUESTIONS

edited by

Robert Rosenberger

Automatic Press ♦ VIP

Automatic Press ♦ $\frac{V}{I}$P

Information on this title: www.vince-inc.com/automatic.html

© Automatic Press / VIP 2010

This publication is in copyright. Subject to statuary exception and to the provisions of relevant collective licensing agreements, no reproduction of any part may take place without the written permission of the publisher.

First published 2010

Printed in the United States of America
and the United Kingdom

ISBN-10   87-92130-34-8   paperback
ISBN-13   978-87-92130-34-1   paperback

The publisher has no responsibilities for
the persistence or accuracy of URLs for external or
third party Internet Web sites referred to in this publication
and does not guarantee that any content on such
Web sites is, or will remain, accurate or appropriate.

Typeset in $\LaTeX 2_\varepsilon$
Cover design by Vincent F. Hendricks

Contents

Preface   iii

Acknowledgements   vii

1 Harry Collins   1

2 John Dupré   13

3 Arthur Fine   27

4 Allan Franklin   33

5 Peter Galison   49

6 Ronald N. Giere   69

7 Adolf Grünbaum   83

8 Sandra Harding   89

9 Don Ihde   103

10 Sheila Jasanoff   109

11 Evelyn Fox Keller   129

12 Philip Kitcher   137

13 Helen Longino   147

14 David Papineau   157

15 Stathis Psillos   171

16 Joseph Rouse   185

17 Patrick Suppes   199

18 Nancy Tuana	221
About the Editor	237
Index	239

Preface

The philosophy of science as a branch of study grapples with a variety of topics. These range from the analysis of the metaphysical underpinnings of scientific claims, to science's place in society; from the articulation of abstract concepts such as "objectivity," "progress," "theory," and "evidence," to the study of concrete historical and contemporary practices; from the description of how science works, to the prescription of how it should. In what follows, eighteen of the most influential figures in contemporary philosophical discussions on science are posed the same five questions. The resulting essays provide thought-provoking and sometimes surprising views on the philosophy of science as a branch of study, on science's methods and claims, and on the societies within which science is practiced.

The questions posed to these figures invite them to relate their personal biographies, and to reflect on their own contributions to the field. The authors are prompted to offer opinions about the most important and exciting topics in the philosophy of science, and about where the field is headed. They are also asked to consider the implications of work in the philosophy of science for the practice of science, for issues of social justice, and for other practical matters. The questions themselves are purposefully broad. This enables the participants to each take her or his answers in an individual direction and to comment on issues relevant to her or his own expertise. I have encouraged these authors to focus on the parts of the five questions to which they have opinions or insights (with several participants even rewriting or reorganizing those questions in the process).

The benefits supplied by this collection of interviews are at least threefold. First, it offers a unique opportunity for engaging the views of these figures. The interview context provides an informal space for these authors to take stock of their own works, and to convey their outlooks on a variety of issues both within and outside of the field. The interviewees provide casual and engaging summaries of their philosophical perspectives, and contextualize these ideas within their personal histories, evolving viewpoints, and their opinions about the field at large. These essays thus provide approachable introductions for readers new to their bodies of work, and helpful supplementation for professionals who make use of their ideas. Each essay is appended with a list of the author's works to facilitate further engagement.

Second, the collection provides a fascinating snapshot of the contemporary field of philosophy of science. A number of overarching themes and explicit concerns emerge. There is significant reflection on the continuing influence of figures such as Quine, Dewey, Wittgenstein, and especially Kuhn, Hempel, and the logical empiricist perspective. Central issues of the field, such as the debate over 'scientific realism,' feminist critique, and the development of specialist discussions of specific sciences (e.g. physics, biology, the social sciences), all receive detailed attention. One issue that arises in several interviews is the question of how philosophers of science should participate in public discussions over how science is demarcated from other discourses, with particular concern over the rise of "intelligent design" challenges to evolutionary theory. Also, a number of thinkers here consider the roles scientists and also philosophers of science—as citizens with certain expertise—should play in public debates over policy and other matters. Taken together, these interviews present a special opportunity for discerning of the contemporary state of the field.

Third, the interviews together present an opportunity to examine the evolving history of the field of philosophy of science. Many of the participants here chronicle personal experiences in relation to past states of the field. These include reflections on encounters with major historical thinkers, on issues of once-burning concern, and on the maturation of frameworks of thought, specializations, and subfields. In addition, a number of the interviewees offer sustained critical analyses of the history of the field, and assess how this history is distorted by the lens of contemporary preoccupations. A thumbnail version of the history of the philosophy of science, it seems, should include at least the following: the dominance of the logical empiricist perspective in the earlier part of the twentieth century; its legendary collapse under the weight of technical philosophical dilemmas; the emergence of anti-positivist alternatives such as Kuhn's work; and the advent of more recent issues such as feminist perspectives, scientific realism, and the sometimes hostile encounters with the emerging field of science studies. Yet in contrast to the tidiness of this thumbnail sketch, the history through which the following interviews navigate is one which is textured, nuanced, and at times less certain. I suspect these accounts will occasion readers working in the philosophy of science to reassess the place of their own work within this evolving history.

The participants of this collection reflect a variety of viewpoints and specializations. Certainly the analytic tradition of philosophy, the conventional residence of the field, is the best represented, though perspectives from the Continental tradition are included as well. As are the views of figures from science studies, a field to which the philosophy of science maintains continual (if at times contentious) interaction—figures important not only for their own contributions to the philosophy of science, but also for their observations of the field itself, posed from the point of view of interlocutors working in a related (if not overlapping) discipline. The interviewees of this volume also reflect a diverse array of specializations and subfields, and they participate in a number of disparate discussions. This collection of interviews as a whole, with its at times contradictory supply of positions, thus speaks to the breadth, sophistication, and complexity of ongoing work in the philosophy of science.

The range of subject matter which receives attention through the course of the collected interviews is extensive. The issues which arise within the following essays include, but are not limited to: the epistemology of experimentation, and the bodily relations users develop with instrumentation; the debate over scientific realism, including analyses of influential arguments such as 'no miracles' and 'inference to the best explanation,' and the relation between realism and scientific practice; and reflections on the notion of 'objectivity,' including its concrete history, its varying usage in the legal systems of different countries, and its relation to epistemological standpoints. The essays explore general issues of public concern, such as how research agendas should be decided, and how scientific expertise should be understood, and specific issues, such as critique of intelligent design, and prescription for the study of climate change. There is significant reflection on the social aspects of scientific practice, such as governing norms and values, subcultures of research and the local languages they develop, large-scale scientific experiments with numerous specialists, and the distribution of cognition across multiple researchers and instruments. Essays consider the experiences of women scientists, the influence of metaphors of femininity and masculinity, the distinction between the notions of sex and gender, and the insights of postcolonial theory. Throughout the interviews, philosophical issues pertaining to a variety of the special sciences arise. For example, on the philosophy of biology, interviewees consider the nature of genes, and the importance of the study of microbes. On the philosophy of physics, issues analyzed include quantum entanglement, and the many worlds theory. On the philosophy of psychology, questions arise over the nature of human consciousness, physicalist conceptions of the mind, the psychology of norms, and critiques of psychoanalysis and of evolutionary psychology. The essays analyze the use of metaphor in science, bayesianism, cognitive accounts of science, and the relation of philosophy of science to sociological and anthropological studies of science. And the list goes on.

As evidenced by the diversity of topics and viewpoints addressed in the following interviews, contemporary philosophy of science can at times appear a series of contradictions; at once a longstanding and well-developed discussion of abstract philosophical topics, and yet at the same time an interdisciplinary field expanding boldly into issues of concrete public concern; a field of study with an established canon, yet with an often misconceived history in need of reexamination and critique; a field both ready to build on previous accomplishments, and to forge paths into new areas of investigation. We are fortunate to have the participants of this volume as guides.

Robert Rosenberger
Georgia Institute of Technology
2010

Acknowledgements

I wish to thank all of those who have helped to make this volume possible. Helpful advice throughout the editing process was provided by Patrick Grim, Vincent F. Hendricks, Sabrina Hom, and Evan Selinger. I also wish to thank the publishers at **Automatic Press ♦ $\frac{V}{I}$-P** for taking on this project. And finally, I am especially grateful for the effort and patience put into this volume by the interview participants.

1
Harry Collins

Distinguished Research Professor

School of Social Sciences, Cardiff University, UK

---

1. How were you initially drawn to philosophical issues regarding science?

I'm a sociologist but for me, the disciplinary boundary between sociology of science and philosophy of science is tenuous. Those who, like me, studied sociology in Britain in the 1960 and 1970s, found themselves dealing with quite a lot of philosophy. Philosophy of the social sciences, and of the sciences, had a big role in the sociology syllabuses of 1960s Britain; I was very much drawn to the philosophical end of things. I like the idea that one could say something general about the type of thing one tried to do as a scientist or social scientist. When I was an undergraduate I found the philosophy of Karl Popper particularly appealing.

A strange thing happened to me shortly after my first degree was awarded. I found myself reading Peter Winch's book *The Idea of a Social Science*. The reason was that there had been a question about it in my final examination and I had never heard of the book and thought I should find out what was in it. So I started to read this book and could not understand a word of it. But I stuck at it for about 3 months and what it was saying slowly dawned on me. It presents itself as a critique of social science—Winch is a Wittgensteinian philosopher who claims that the deep questions of sociology are misbegotten philosophy. But there was something fascinating about the book—it was a kind of thinking that was completely new to me. Being intrigued by Winch I also tried reading Wittgenstein's *Philosophical Investigations* and it seemed to me that I understood it pretty well—after struggling to extract the sense from Winch it was not so hard to see what Wittgenstein was getting at; at least I got the Winchian interpretation and that has always worked well enough for what I want to do.

A bit later I picked Thomas Kuhn's *Structure of Scientific Revolutions* from the shelf in a bookshop – I'd never heard of it but the title sounded interesting. Kuhn seemed absolutely straightforward – a direct application of Winchian/Wittgensteinian ideas to science. In fact if you read around page 120 in Winch's book (which was published in 1958) you will find the same idea as Kuhn would refer to in 1962 as a 'paradigm.' Winch talks of the germ theory of disease as involving a whole new way of acting in the world as well as being a new set of concepts—concepts and ideas are treated as two sides of the same coin. Years later I was to find that Kuhn was not much influenced by Wittgenstein but I believe that if he had been he would have understood his own idea of paradigm better and would not have been so willing to pull it apart into separate practical and conceptual components.

Somewhere along the line I realized that though Winch was an attack on the idea of sociology, he could be construed as actually doing sociology. What he was saying was that the world of practice and the world of concepts could only be understood together. This means that if you want to study concepts you can study practice or if you want to understand practice you can study concepts so sociology and philosophy become one. Nowadays when I read Wittgenstein it seems to me I am reading a sociologist making observations about how people live in societies. Admittedly he doesn't use what are usually thought of as sociological methods but his remarks are based on observations of how people act in the ordinary way—he is, as one might say, the quintessential participant observer of our own society.

The permeability of the boundary between sociology and philosophy has been made concrete for me by the fact I have co-authored a book with a philosopher and that I have published in philosophy journals and am occasionally asked to present my work at philosophy conferences. It has to be admitted, however, that in many ways I am no kind of philosopher at all. For example, I do not know the philosophical literature in the detailed way that professional philosophers know it and I have little interest the analysis and re-analysis of philosophical texts so that we come to know what 'X really meant.' For example, it doesn't matter to me if Wittgenstein 'really' meant something different to the interpretation that I have wound up with. I am interested in the extent to which philosophy can tell me whether the world works 'this way' rather than 'that way' and the interpretation of Wittgenstein that I have fits with just about everything else that I know so that's

OK. Philosophy for me is a bit closer to the sciences than to the humanities. Furthermore, for the most traditional philosophers of science my practice might seem subversive as I want to replace the old quasi-logical style of analysis with what is likely to be read by traditionalists as a sociological perspective. In particular, I start with the idea that the prime location of knowledge is the collectivity rather than the individual. To repeat, if one takes Wittgenstein seriously this seems philosophical but if you don't it might seem imperialistically sociological.

2. What, in your view, are the most interesting, important, or pressing problems in contemporary philosophy of science?

Well, this is even more subversive. It seems to me that over the last three decades the socio-philosophical approach has been so successful that the traditional problems of the philosophy of science no longer seem so pressing. We now know that science creates its knowledge in much the same way as any other cultural endeavour; the fact that science rests on theories and experiments does not make any really deep *epistemological* difference. Theorizing and experimentation are always going to be central to science as a practice but the closer the examination of that practice the less does it appear to depart from ordinary reasoning—at least, it does not seem to depart in any deeply principled way which makes it, as opposed to ordinary reasoning, the key to true knowledge. There is still valuable work to be done in showing how the practice goes on but we now know how to do the studies and get the answers and it is what any PhD student is expected to be able to do in a fairly routine way.

I am still doing work of this kind but nowadays it is more driven by my wanting to complete what is coming up to being a four-decade long study of the detection gravitational waves rather than a desire to work out the solutions to deep conceptual problems. What I am doing is filling out details—trying to document the ways that the scientists will finally decide that some proto-gravitational wave signal counts as a real gravitational wave signal rather than a piece of noise. On the half-dozen or so previous occasions that proto-signals have turned up they have always decided they were noise—I want to be able to say what the difference is, if there is a difference, when they finally decide the have found something.

This does not mean that life is not exciting, however, quite the contrary. I think life for the socio-philosopher of science is more exciting now than it has been at any time since the early 1970s when the sociological turn began, but this is because a new kind of question has to be asked.

If I can use the language of a now notorious paper published, in 2002, by myself and Robert Evans, there have been two 'waves' of science studies and we now need a third wave. The first wave started with the premise that science was something special and philosophers tried to explain how it worked while sociologists, like Robert Merton, tried to show the social conditions that made for its existence—the famous 'norms of science.' The second wave, which began with the sociological turn and became absorbed (to what I believe was its detriment) in the general movement called post-modernism, showed that science was not so special after all. It showed that there were no special methods nor norms with epistemological significance. There were just lots of different ways of doing what we call science and there was no 'philosopher's stone' that turned the base metal of ordinary thinking and argument into the gold of true science. In my opinion the second wave killed the first wave as a pioneering intellectual endeavour, though academic institutions in this situation tend to maintain momentum by looking more inward than outward—like a spinning skater pulling in their arms.

The new third wave does not kill the second wave—which runs on as strongly as ever in theory if a little predictably in practice—but the third wave asks what we should do with the insights of the second wave. The second wave provides no guidance to policy makers who have to think about how to use what are now understood to be the insecure findings of science; the third wave tries to fill this gap. The problem presents a very difficult and fascinating set of questions which require new kinds of answer.

The new questions grow on very shallow philosophical roots— the way we prefer to live our lives. The roots are about as shallow as the roots of ethics. Somehow, in spite of everything that has been discovered by Wave 2, we still have a preference for the methods and findings of science even if we cannot explain why— just as we cannot really explain why we have a preference for not gratuitously torturing children or whatever (ironically, this kind of 'preference,' or certainty, may be the most profound of all). No academic who thinks hard about the consequences of abandoning science can relish the prospect of living in the society that would result. What would a society be like that did not give extra weight to the opinions of those who formed their views with the aid of study, observation and experience? (Just for a start, there would be no jobs for academics!) Wave 3 has to work out how to describe the kind of society that we do want, and the meaning and role of science within it. Now that Wave 2 has destroyed the comfort and security of the Wave 1 model of science this becomes a new kind of task and a very demanding one.

The approach that my colleagues and I have adopted is to ask how a policy-maker should act in the situation before there is any settled scientific consensus over a controversial matter. For example, how should policy-makers make their choices between a range of medical treatments when there are disputes such as are found in certain cases of vaccination or AIDS remedies? Wave 2 shows us that we aren't going to find the answer by looking for the undisputed truth of the matter because the undisputed truth, if it ever emerges, won't emerge until long after the policy decision has to be made. You can't wait for scientific truth when you make policy because in the meantime people might be dying. As we put it in *Rethinking Expertise*, the speed of politics is faster than the speed of science. The approach we adopt is to forget about the truth of the matter and just go for a potentially fallible analysis of who has expertise. We simply say, let us allow all those who 'know what they are talking about' to contribute more to the decision than those who don't; they may get it wrong but it is the best that can be done.

The hard job starts with working out who can be taken to 'know what they are talking about' and exactly what it is that is being talked about. We begin with a categorization of types of expertise which we call the 'Periodic Table of Expertises.' It can be found in our book *Rethinking Expertise* or on our website (www.cf.ac.uk/socsi/expertise or Google 'Harry Collins Expertise' and look for All@SEE).

But this is just a 'taste' of what I think is the new and exciting set of questions. The true role of science is, I believe, only just being understood (though the spirit if not the substance, I have been reminded, can be found in Karl Popper's *Open Society and its Enemies*). The role of the 'norms of science,' including those discussed by Merton, is as a moral foundation to life. A 'form of life' that is based on a preference for free criticism with a view to reaching a collective truth and based on observation and experience as a justification for an opinion is a good in itself. One might think of the new role of the philosopher as jumping off from the analysis of Hobbes and Boyle found in the book *Leviathan and the Air-Pump* by Shapin and Schaffer. Hobbes feared that science allowed ordinary people (strangely it is the scientist that fills the role of the ordinary person in that book), to question the authority of the monarch and generate disorder. Shapin and Schaffer argue that science, being messy and untidy, itself requires the equivalent of monarchs to bring it to order—experiments and theories alone will not do it. They conclude that 'Hobbes was right.' And Hobbes was right when it comes to accurate description of how science works. But when it comes to how we want to live our lives today, Boyle was right. We want to make our decisions on the basis of observation and experiment, not authority.

The philosopher of science must become more like a moral philosopher, working out the 'ought's' of how we want to live rather than the 'is's' of how we do live. The job of the philosopher of science has to be to work out how to describe the good, what we take to be good about it, and how it is to be demarcated from the not-good. Some of the substantive content was there in Wave 1 but the whole thing has to be turned on its head. Whereas Merton tried to justify democratic norms by showing their epistemological potency—it was these norms that made Western democracies so scientifically and technologically successful—we now have to try to understand the norms as goods in themselves. We now know that we cannot base the norms on their epistemological or technological potency because science and technology can, and often does, get along without them. It just happens that the scientific community invented the norms of the good life as their justificatory myth.

With the death of Wave 1 the new problems are of a really interesting and new kind. What is the difference between science and the arts? How do you demarcate the idea of 'intelligent design' from the theory of evolution? (And remember, there are experts in all camps.) Perhaps what is found in the last Chapter of *Rethinking Expertise* offers the beginning of an idea about how to think about these things, perhaps not: I expect better philosophers will be able to do better so long as they don't fall back into the old quasi-logical Wave 1 way of thinking. The answers have to be found in the way we want to live our lives, not the most efficient way to get to the truth of the matter.

3. How has your work offered original contributions to discussion on science? What does your work reveal that others fail to appreciate?

This is a question inviting a display of hubris. I could list about a dozen distinctive concepts that I think I have invented—for example, a nice neat one is the distinction between evidential individualism and evidential collectivism. I guess the one for which I am best known is the 'experimenter's regress.' This shows why experimental replication has to be a social process involving mundane decisions and cannot be an automatic method for uncovering truths about the world. The idea of which I am most proud is the distinction between mimeomorphic and polimorphic actions—most carefully worked out in the book called *The Shape of Actions*. This is the least well-known aspect of my work, and quite a tough one to use, but if someone told me that just one of the ideas associated with my name will still be influential in 50 years that's where I would put my money.

The most recent set of ideas turns around the 3rd Wave and, in particular, includes the new notion of 'interactional expertise.' This is all so obvious once you get the hang of it that once the current, and astonishingly strong, resistance to it dies, it will just become commonsense. I have just completed a book about tacit knowledge and dividing it up in new ways which, for me anyway, help me understand better what it means. One concept in the new book (already set out in a paper called 'Bicycling on the Moon,') is 'somatic-limit tacit knowledge.' It is the idea that some things—such as Polanyi's classic example of balancing on a bike—are not tacit at all since the physics can be described. We think of them as tacit knowledge because we cannot use the rules due to the limits of our bodies and brains. If our brains worked a million times faster (or we were cycling on the Moon or a small asteroid so the bike fell more slowly) we would be able to use the explicit rules and we wouldn't be tempted to call bike-balancing ability tacit knowledge. I think the analysis of tacit knowledge has been ill-served by the concentration on the abilities of the body and the crucial feature is knowledge embedded in social groups—this goes back to polimorphic and mimeomorphic actions.

4. What is the relation between philosophy of science and scientific practice, science policy, or efforts for social justice? Can there be a more productive relation? Is this desirable?

Richard Feynman is said to have said that scientists need philosophy of science like birds need ornithology. To some extent he was right but there is an overlooked corollary, namely that if you want to know about ornithology you shouldn't ask birds.

I think Wave 2 doesn't have a lot to say to science except that the softer-edged model it gives rise to could lead to more non-traditional people taking up science and to less disappointed PhD students—disappointed (and sometimes accounting themselves failures) when they discover that the quasi-logical set of procedures they had been taught about as undergraduates bears no resemblance to what happens at the research front. Of much more importance is that scientists are also citizens and in that role, when they are asked to comment in public on the role of science, they need to have their justificatory myths replaced by the much more fallible picture of science that comes with Wave 2 and Wave 3.

As I have argued, the real pressing and immediate problem for science is to raise its ways of going on to a position of moral preeminence. This isn't a matter of Richard Dawkins-type harangues about the superiority of scientific truth over religion—something that gives atheism a bad name—but a careful and modest exploration of the inherent values of observation, experience and free debate over the values of revelation or the exercise of power and authority. Unfortunately, the dominant position in science studies these days is the Machiavellian model championed by Bruno Latour in which science is just politics by other means and scientific truth making is best achieved by Machiavellian maneuvers. To advocate these models is to destroy the idea of science as a moral good. The old Wave 1 scientific myths—the single individual having the right to try to stand against the collective opinion and insist that the emperor has no clothes—is the kind of myth we need to inform a good society however bad that is as a description of scientific life. And, of course, now that it has served its political purpose of de-legitimating the essentialising of race and gender, we need to get past the distrust of dualities, with its treatment of humans, with their rich languages and social life, as the same as things and animals. For all its academic success and attractiveness, just a few moments of reflective observation shows that such a position is no more than a philosophical conceit—that emperor is naked. One of the most remarkable spectacles to be seen today is the lemming-like rush of sociologists to abandon the very notion that human society has special qualities—without which notion there can be no distinctive subject called sociology (nor any hope of understanding tacit knowledge).

5. Where do you see the field of philosophy of science to be headed? What are the prospects for progress regarding the issues you take to be most important?

I think the prospects are marvelous and the future looks fascinating but it is a future in which what is traditionally thought of as philosophy of science plays a less central role. A remodeled philosophy of science will explain and explore the nature of a society informed by the traditional myths of science. It will elevate the norm that observation and experience should be a basis for opinion to the centre of what we count as a good society. But it will do it without epistemology.

References

Kuhn, T. *The Structure of Scientific Revolutions.* University of Chicago Press, 1962.

Popper. K. *The Open Society and its Enemies.* London: Routledge, 1945.

Shapin, S., and S. Schaffer. *Leviathan and the Air Pump: Hobbes, Boyle and the Experimenal Life.* Princeton: Princeton University Press, 1987.

Winch, P. G. *The Idea of a Social Science.* London: Routledge and Kegan Paul, 1958.

Wittgenstein, L. *Philosophical Investigations.* Oxford: Blackwell, 1953.

Selected Bibliography

Authored Books

Frames of Meaning: The Social Construction of Extraordinary Science. (with T. J. Pinch) Henley-on Thames: Routledge and Kegan Paul, 1982.

Changing Order: Replication and Induction in Scientific Practice. Beverley Hills & London: Sage, 1985; 2nd edition, Chicago: University of Chicago Press, 1992.

Artificial Experts: Social Knowledge and Intelligent Machines. Cambridge, Mass: MIT press, 1990.

The Golem: What Everyone Should Know About Science. (with T. J. Pinch) Cambridge & New York: Cambridge University Press, 1993; new edition, 1998.

The Golem at Large: What You Should Know About Technology. (with T. J. Pinch) Cambridge & New York: Cambridge University Press, 1998.

The Shape of Actions: What Humans and Machines Can Do. (with M. Kusch) Cambridge, Mass: MIT Press, 1998.

Gravity's Shadow: The Search for Gravitational Waves. Chicago: University of Chicago Press, 2004.

Dr Golem: How to Think About Medicine. (with T. J. Pinch) Chicago: University of Chicago Press, 2005.

Rethinking Expertise. (with Robert Evans) Chicago: University of Chicago Press, 2007.

Gravity's Ghost: The Equinox Event and Science in the 21st Century. Chicago: University of Chicago Press, forthcoming, 2010.

Tacit and Explicit Knowledge. Chicago: University of Chicago Press, forthcoming, 2010.

Edited Books

Knowledge and Controversy: Studies in Modern Natural Science, Special Issue of Social Studies of Science. 11, 1, Beverley Hills & London: Sage, 1981. (Based on 1980 Bath conference).

The sociology of Scientific Knowledge: A Sourcebook. Bath: Bath University Press, 1982.

Humans, Animals and Machines: Special issue of Science Technology and Human Values. 23(4): 371-490. Co-edited with M. Lynch. Beverley Hills: Sage, 1998. (Based on 1995 Bath conference).

The One Culture?: A Conversation about Science. Co-edited with J. Labinger. Chicago: University of Chicago Press, 2001.

Case Studies in Expertise and Experience: Special issue of Studies in History and Philosophy of Science, 38(4), 2007.

Selected Articles

"The TEA Set: Tacit Knowledge and Scientific Networks." *Science Studies.* 4: 165-186, 1974.

"The Seven Sexes: A Study in the sociology of a Phenomenon, or The Replication of Experiments in Physics." *Sociology*. 9(2): 205-224, 1975.

"Socialness and the Undersocialised Conception of Society." *Science, Technology and Human Values*. 23(4): 494-516, 1998.

"The Meaning of Data: Open and Closed Evidential Cultures in the Search for Gravitational Waves." *American Journal of Sociology*. 104(2): 293-337, 1998.

"Tacit Knowledge, Trust, and the Q of Sapphire." *Social Studies of Science*. 31(1): 71-85, 2001.

"The Third Wave of Science Studies: Studies of Expertise and Experience." (with R. Evans) *Social Studies of Science*. 32(2): 235-296, 2002.

"Interactional Expertise as a Third Kind of Knowledge." *Phenomenology and the Cognitive Sciences*. 3(2): 125-143, 2004.

"Bicycling on the Moon: Collective Tacit Knowledge and Somatic-Limit Tacit Knowledge." *Organization Studies*. 28(2): 257-262, 2007.

"They Give You the Keys and Say 'Drive It:' Managers, Referred Expertise, and Other Expertises."(with G. Sanders) *Studies in History and Philosophy of Science*. 38(4): 621-641, 2007.

"Mathematical Understanding and the Physical Sciences." *Studies in History and Philosophy of Science*. 38(4): 667-685, 2007.

2

John Dupré

ESRC Centre for Genomics in Society (Egenis) and Department of Sociology and Philosophy

University of Exeter, UK

---

1. How were you initially drawn to philosophical issues regarding science?

Among my main childhood pursuits were chemistry—bangs, stinks, and suchlike—and natural history, mainly turning over stones and sometimes killing whatever was under them and pinning it on a piece of cardboard. At school I naturally specialised in the sciences and imagined that I would continue with these to university and beyond. Towards the end of my time at school I began to realise that I lacked the patience and the care to be a good scientific experimenter and I was fortunate enough to have the chance of some exposure to philosophy. The confluence of a growing fascination with the practice of philosophy and the increasing awareness that I was not cut out for the scientific coalface made the philosophy of science an increasingly inevitable destination.

My initial point of entry into the philosophy of science was also very much affected by parallel more or less scientific interests. By my college days my natural history interests had moved from creepy-crawlies to plants, and I spent a good deal of time attempting to relate wild plants to pictures and descriptions in various field guides. I realised that arranging objects into kinds was both an important practice in biology and a perennial problem in philosophy, and came to see that many of the assumptions that contemporary philosophers were inclined to make about the nature of this problem fitted very poorly with my practical experience of botany. Philosophers were writing about the problem of natural kinds and the meaning of natural kind terms as if it were self-evident that life provided us with a great array of distinct kinds, each distinguished by some unique essential property. But biology as I encountered it, and also as any engagement with evolutionary theory should lead one to expect, was a much messier business. My first published paper (1981) explored this conflict.

This project provided me with two convictions that have been important for much of my subsequent philosophical career. The first was that the central questions in philosophy could not be addressed without engagement with the deliverances of the natural sciences. I thus allied myself with some, at least, of the contemporary philosophers describing themselves as naturalists. But the second was that a lot of such engagement, even by many of those who embraced this designation, was in practice with caricatured versions of science, or the science of a couple of centuries ago. So the second conviction was that engagement with science that did not attempt to come to grips with the real science was often worse than useless. James Ladyman and Don Ross (2007) have recently described the use by contemporary metaphysicians of the 'A-Level Chemistry' model of physical science, referring to the billiard ball/solar system image of the atom still presented uncritically to British high school students. In just the same way, some contemporary philosophers (e.g. Dennett, 1995) like to appeal to a classical mid-twentieth century model of evolution and imagine that this can provide profound insights into the nature of human life. This has not, in my view, been a productive activity.

Ultimately my interests since I made the decision to study philosophy have been in the big philosophical questions of metaphysics, epistemology, and ethics. But I am also firmly convinced that it is to say the least perverse to approach these without attempting to understand our best grounded empirical accounts of the world. In retrospect this conviction again makes work in the philosophy of science, or on the borders between philosophy of science and those traditionally central areas of philosophy, seem inevitable for me.

2. What, in your view, are the most interesting, important, or pressing problems in contemporary philosophy of science?

Following from what I have just said, I am interested in some quite basic questions about what philosophy of science is, how it should be done, and how it relates to other areas of philosophy that sometimes seem to offer competing approaches to the questions it addresses. Perhaps the most basic such question is, to be slightly paradoxical, whether there are any basic questions in philosophy of science. Without the paradox, are there really questions about *science*, as opposed to questions about particular *sciences*. This connects with an issue with which I have been concerned for many years, the unity or disunity of the sciences. Since I am very sceptical of their being any simple feature that makes a practice scientific or non-scientific, it is difficult to know what a philosophy of science, *tout court*, should be about. I do think that there are epistemic virtues, and that practices that have more or fewer of these can be described as more or less scientific. But it is difficult to see how one could identify such virtues other than by reflection on their use in particular practices and the contribution their use makes to the success of such practices. Probably there are rules of argument or rationality the sufficiently general violation of which makes a practice non-scientific, but this doesn't tell us much about what science positively is.

So I believe that philosophy of science needs to pay more attention to its relation to the rest of philosophy, and perhaps sometimes assert more aggressively its relevance to questions often supposed to be none of its business. One naturally thinks, for instance, of the question of knowledge. In the UK, where I now work, philosophy of science is a peripheral area of philosophy (philosophy of biology is the periphery of a periphery) and the traditional problem of knowledge, or epistemology, remains at the centre. Broadly speaking, at the centre, the project of giving a philosophical account of knowledge is still quite widely seen as that of discerning the conditions under which the statement "S knows that p" is true. This, in turn, is to be accomplished by consulting one's intuitions about how various situations, often bizarre situations, should be described. (I take myself to know that Mary is in Paris, for which belief I have excellent grounds. I infer that she is in Paris or Rome. In fact she is in Rome. Do I know that she is in Paris or Rome?)

There are many things about this procedure that might worry us. Where do our intuitions come from and what do we do if other people have different ones? Should we anyhow trust intuitions about strange and unfamiliar situations? Even if there is a concept we can reach by such intuition, is it necessarily the one that we care about philosophically? And so on. But philosophers of science should have a different kind of worry. They, and also historians and philosophers of science, who explore in detail the processes by which hard won empirical knowledge is obtained from Nature, cannot help wondering what the relevance of traditional philosophical accounts of knowledge is to the processes they explore. To many, the answer seems to be: None at all. It may, further, seem that the right way to explore the nature of knowledge is to explore in detail the processes by which it is established, contested, and so on. Having a lot of sympathy with such views I think that philosophers of science should be more forceful in contesting their sometimes marginal status in philosophy. Of course, deciding what knowledge is on the basis of intuitions about cartoon science, such as A-Level chemistry (as mentioned above), is probably a good way of getting the worst of all worlds.

Turning specifically to my own field of philosophy of biology, the central current challenge, in my view, is the necessity of coming to terms with an explosion of knowledge in molecular biology, broadly construed, that has begun to challenge most of the conventional wisdom in biology generally. This has, arguably, been one of the most remarkable and rapid explosions of knowledge and technical capacity in the history of science, and merely understanding this historical phenomenon better should provide plenty of work for philosophers and historians of science. This in itself is something of an institutional challenge. Philosophers of biology, for the few decades over which it has emerged as a distinct branch of philosophy, have been obsessed with evolution. This has surely received more philosophical attention than all other branches of biology put together. Yet work on evolution is really quite a minor part of contemporary biology, certainly compared with the vast efforts going into molecular biology and projects such as systems biology and synthetic biology currently emerging from molecular biology. Even if, as is still often said, quoting Theodosius Dobzhansky, nothing makes sense in biology except in the light of evolution, this is not where the major action is in the science. There are exciting things happening in evolutionary theory, but these are largely the consequences of our view of evolution attempting to react to the remarkable findings from molecular biology. So, for example, much of the twentieth century history of evolutionary thought, culminating in Richard Dawkins' (1976) well-known image of evolution as the selection of genes, has assumed implicitly or otherwise that the adult organism was somehow inscribed in the genes, and evolutionary processes could be understood without worrying about the process of development or, as has received growing recent attention, the two-way interaction between organism and environment (Odling Smee, Laland, and Feldman, 2003). But it is now quite clear that genomes don't work this way (Barnes and Dupré, 2008), and the organism is no more inscribed in the DNA than it is in the blood, or the cell membrane. Contemporary thinking has become far more open about possible evolutionary processes and even the ultimate taboo, so-called Lamarckianism, or the inheritance of acquired characteristics, is back on the table. But what is also clear is that that insight into evolution will not come from a priori reflection, or cunning work in the theory of games, but will depend on coming to terms with the fine structure of the biology.

I would like to mention one particularly exciting development in biology that has been a direct consequence of the rise of the tools and methods of molecular biology. This is the gradually dawning awareness of the importance of microbial life. Microbes, single-celled organisms, are overwhelmingly the dominant form of life. For 80% of the history of life they were the only life-forms, and even today they comprise more than half the total biomass on Earth (if one excludes the extracellular material in plants). Complex multicellular life forms (macrobes), moreover, are entirely dependent on symbiotic relations with large and diverse communities of microbes.

Philosophy of biology (and a lot of theoretical biology), on the other hand, has been almost entirely concerned with macrobes. This has led to the assertion of general theses about life that in fact apply only to this unique and exceptional area of biological diversity (O'Malley and Dupré, 2007). One central example illustrates the extent of the problem. Many philosophers and biologists still propose the so-called Biological Species Concept (BSC) as the central account of what a species is. According to the BSC, a species is a reproductively isolated and interbreeding group of organisms. But this concept has no obvious application to asexual organisms (i.e., microbes), since these do not form interbreeding groups. More interesting still, however, has been the growing realisation that many microbes employ several mechanisms for the exchange of genetic material. This genetic exchange, moreover, is by no means limited to closely related organisms. Thus to the extent that microbial groups might, on the basis of this genetic exchange, be counted as interbreeding, they are far from reproductively isolated. These phenomena begin to threaten an image that underlies most post-darwinian thought about both evolution and biological taxonomy, the tree of life. The tree of life is a branching structure intended to represent the history of the divergences between different kinds of organisms. Since it is assumed that branches once separated remain distinct, it implies that all the ancestors of an organism on a particular branch of the tree can be found by moving downwards towards the trunk of the tree. The lateral gene transfer (LGT), found to be common among microbes, contradicts this picture. Some microbiologists (e.g. Doolittle, 2005) are now urging that we should replace the image of the tree by one of a network, at least for the case of microbes, which is to say for most of the history of life. This will require some fundamental rethinking of the issues that have most concerned philosophers of biology for the past thirty years.

3. How has your work offered original contributions to discussion on science? What does your work reveal that others fail to appreciate?

My philosophical work can be divided roughly into three stages. The first, culminating with my book *The Disorder of Things* in 1993, was concerned with attempting to develop an alternative to what was still a standard set of views about science that seemed to me inadequate to capture the distinctive nature of biology. Philosophy of biology did not exist as an identifiable subfield of philosophy until the 1970s, in part because it was widely assumed that the only fundamentally interesting science was physics (this view is by no means extinct today). Since the field emerged, it can be seen as gradually asserting its independent and distinctive nature and, as it did so, making a central contribution to growing scepticism about the vision of science as a unified whole that underlay the assumed primacy of physics. This was the overall project with which this stage of my career was deeply involved.

The first area in which I explored the distinctive character of biology was in relation to a view about natural kinds derived from reflection on physics and chemistry. This view sees one of the main goals of science as that of telling us the right way of classifying the elements of nature (the natural kinds). I became convinced at an early stage that there was no such correct way of classifying biological things. Different questions or interests will often require different principles of classification. I was, however, concerned to avoid a natural extension of this thought to the position that any way of classifying things was as good as any other. Given a particular interest, nature often provides us with the right way of classifying things. The combination of a pluralism about classification with the modest realism involved in blocking the descent into pure nominalism, was the position I have referred to for many years as promiscuous realism. Although debates about the 'correct' species concept still carry on to some extent in biology, pluralism is becoming something of a consensus among philosophers of biology. The insight into microbial life made possible by the tools of molecular biology should cement this development.

My most central concern in this period, however, was with the doctrine that provided the foundations for the perception of physics as the fundamental science, reductionism. According to classical reductionism, ultimately our scientific understanding of complex objects should be derivable from our knowledge of their fundamental constituents. Perhaps the details of such derivation will go beyond our limited powers and even beyond the capacity of our best computers, but that is a practical limitation not a theoretical one. Much of my early work was concerned with developing the hunch that this kind of reductionism was not merely unachievable in practice, but fundamentally misguided. The unachievability in principle has become something like a commonplace, certainly with respect to biology. However the underlying intuition, that in the end there is nothing but physical stuff (in the famous words of Ernest Rutherford, "All science is either physics or stamp collecting") remains quite widely held. Surprisingly, perhaps, I take my most recent work, on molecular biology, to confirm my hunch that this reductionist perspective is fundamentally misguided.

The second phase of my philosophical career, occupying much of the '90s, was driven by the feeling that philosophy of science should serve a critical role, and should have an important role in exposing the dubious epistemic credentials of questionable claimants to the mantle of science. In this respect, I find myself in uncharacteristic sympathy with perhaps the most influential philosopher of science of the 20th century, Karl Popper. Unlike Popper, however, I do not believe there is any unique criterion that can be used to sort the scientific sheep from the goats. Indeed, my commitment to the disunity of science precludes any such criterion. Part of the motivation for this work, then, was the wish to demonstrate that scientific pluralism did not entail unrestricted tolerance: molecular genetics, for instance, was good science, intelligent design theory wasn't. In the *Disorder of Things* I had proposed that this distinction should be grounded on the identification of a range of epistemic virtues and, broadly speaking, the more epistemic virtue a project had, the better its claim to be part of science.

The project whose virtue I was mainly concerned to call into question was evolutionary psychology, the modern successor to the socio-biology made famous by E. O. Wilson (1975) and subsequently infamous through a series of devastating critiques (e.g. Kitcher, 1985). My diagnosis of the defects of evolutionary psychology suggested a quite general pathology, what I called Scientific Imperialism, the attempt to deploy a successful scientific idea too far beyond the domain in which it has been successful. So while no one serious doubts that evolution has been a productive and successful theory, the assumption that it has the resources to provide significant insight into human behaviour has yet to find much confirmation. And in fact I argued that there were good reasons to doubt whether it will ever do so. It is plausible that imperialism in this sense may be an almost unavoidable feature of scientific development: if a scientific idea is successful it is quite natural to see how far it can be pushed. And if this is right, it suggests a general area where philosophical critique should be useful, the exploration of the scope and limits of central scientific ideas.

In the last few years I have been almost entirely concerned with biological insights that have emerged from recent molecular biology, especially genomics. This has led me to the interest in microbes mentioned above, in the emerging fields of systems and synthetic biology, and ultimately in reconceptions of basic biological concepts, such as organism or gene, and the development of relatively new concepts, such as biological system, that these advances have necessitated.

The point I would like to stress most strongly about this work is more methodological than systematic. I have been very fortunate to have received major funding from two UK Government research funding agencies, from the Economic and Social Research Council to establish the institute in which I work, and subsequently from the Arts and Humanities Research Council to develop specifically the programme of work in philosophy of biology. This has enabled me to assemble a group of about 30 researchers and PhD students with philosophical, sociological, or historical interests in the biological sciences, and especially in recent molecular-based biosciences. I mention this not (merely!) to promote and express gratitude to my sponsors, but rather because I am convinced that this is the right way to work in this area. Contemporary biology is too big and fast-moving for an isolated researcher to engage with on anything but a very narrow front. If philosophy of science is to engage effectively with the important social issues on which it has an increasingly important bearing, it needs to attempt a far more inclusive view of the science. This requires collaboration between researchers engaged with different parts of the unfolding scientific terrain, and also a range of academic disciplines—philosophers, social scientists, historians, and experienced biologists.

The professional ethos of philosophy is very much of the single isolated researcher, and this will no doubt be hard to change. Of course I do not mean to suggest that isolated philosophers will not continue to do important work; it is rather that certain kinds of work will not be done unless some philosophers adapt to a more collaborative kind of activity. It is striking that this suggestion reflects developments in both the practice and the content of biology. On the practice side, one of the main directions in which biology is currently moving is the promotion of systems biology. Systems biology is, very roughly, the attempt to integrate the massive quantities of data being generated by high throughput molecular techniques into some kind of understanding of the larger systems in which the molecules are involved (O'Malley and Dupré, 2005). Much about this field remains controversial, but what is uncontroversial is that it will require collaboration between, at least, 'wet' biologists, computer scientists, and experts on mathematical modelling. Some systems biologists have even embraced collaboration with philosophers. As sociologists of science have observed, the development of systems biology implies new working relations for many of the scientists involved, and sometimes clashes of disciplinary culture.

In terms of the object of study, what I had in mind is the increasing awareness of the importance of symbiosis. 90% of the cells in a human body are not the cells that were sequenced in the Human Genome Project, but microbial symbionts, especially in the gut, but also on the surface of the skin and in every body cavity. It has become increasingly clear that these are not mere opportunists exploiting a vacant niche, but that many are essential for the proper functioning of the human organism. These 'non-human' parts of the human, because of their diversity, also contain perhaps 99% of the genes in the human body. Reflections on such phenomena, and there is nothing exceptional about the human case, lead to the thought that we need to rethink what an organism is. If it is the whole that functions in interaction with the environment, then it should be equated with the full system of interacting cell-types; and our conception of a multicellular organism as a single genomically homogeneous cell-lineage must be abandoned. In recent work I have advocated the general view of an organism as a collaborating assembly of segments of diverse cell lineages (Dupré and O'Malley, 2009). Such a definition, incidentally, allows us to see the complex communities in which microbes generally live (such as biofilms) as a kind of organism, thus providing an interesting conceptual unity across diverse parts of biology.

If indeed biology is becoming the collaborative multidisciplinary study of collaborative multispecific entities, it may not be entirely fanciful to suppose that the best way to study it might be in the context of collaborative and diverse academic groupings. At any rate, this is a thesis that I and my colleagues are attempting to illustrate practically at the present time, and I hope we have already had some success in doing so.

4. What is the relation between philosophy of science and scientific practice, science policy, or efforts for social justice? Can there be a more productive relation? Is this desirable?

My view on this question is implicit in much of what I have said before. Although a very few philosophers who have developed general accounts of 'science,' notably Popper and Thomas Kuhn, have had considerable impact on science, it is very doubtful whether the influence has been healthy. The central ideas of these philosophers have often not been well understood, and have generally been used as blunt weapons to attack opposing scientific ideas or assert the impeccable credentials of one's own. I have little doubt that serious impact on scientific practice will depend on detailed engagement with specific scientific developments and with scientific practitioners. Philosophy of biology has a very good track record in this respect, with many leading figures having collaborated to varying extents with practicing scientists. Despite the declining tradition of seeing science as a quest for laws, biologists are much more inclined to talk about concepts as the central organising resources in their field. Analysis and critique of concepts is arguably what philosophers are best at, so there seems reason to be sanguine about the prospects for contributions to the core issues in biology from philosophy.

Science policy and, perhaps more important, scientifically informed policy, are another matter. Access to policy-makers is generally difficult, and academics who make serious efforts to achieve such access may find they have little time left for research. Again drawing on my own experience, I would say that the social organisation of academic work is a crucial issue. Academic departments are not generally much concerned with giving advice to policymakers, and they are not places that policymakers typically look to when they feel the need for specialist advice. Focused research institutes on the other hand tend to be different in both these respects. When they are funded by governments they will often be funded precisely with the view that they will be useful sources of practical advice, and since giving advice is therefore often one of their indicators of success they will often be set up to look for opportunities to offer it. To the extent that they are much more specifically focused on particular issues or clusters of issues they can be reliable places to look for certain kinds of expertise. If philosophers of science have useful contributions to make on issues of social justice, their ability to do so will be much enhanced by participation in institutional structures within which advice-giving is an important ambition.

Of course none of this addresses the question whether these are appropriate ambitions for philosophers of science. I happen to believe that philosophers thoroughly versed in particular areas of science, are often very well-equipped to provide policy input deriving from this expertise. Philosophers are often good at seeing consequences, formulating arguments, and suchlike, which seem to be central parts of the appropriate skill set. But of course they may prefer to do other worthwhile things, such as write books or teach students. There is another point, however. In the UK, at least, practical relevance is increasingly becoming a sine qua non of government funding of research. However much some philosophers may deplore this development, it looks increasingly to be a fact we shall have to live with. It should be cause for some satisfaction, then, that philosophy of science is a discipline that has great potential for thriving in such a situation. Science is, after all, one of the central organising institutions of contemporary societies, and its nature and products are often very poorly understood by most members of that society. As experts in the interpretation, analysis and critique of science it should not be difficult for philosophers of science to claim a very practical social role.

5. Where do you see the field of philosophy of science to be headed? What are the prospects for progress regarding the issues you take to be most important?

This leads me very naturally to the final question, where is philosophy of science heading? As I have already indicated, I anticipate that philosophy of science, or anyhow philosophy of biology, will continue to increase its close engagement with developing science, and also its engagement with the wider social issues that emerge from, or are affected by, developments in science. I think this development will be greatly enhanced if, as I suspect, philosophers increasingly work within transdisciplinary institutions that promote contact with scientists, other areas of science studies, and other philosophers with related interests. I would entirely welcome this development.

Will philosophy of science solve the main problems it currently faces? A problem for this question is that philosophy of science faces an ever-faster moving target. Like the Red Queen, it will do very well if it manages to stay in the same place. The ambition, I think, if philosophy of science accepts that its role is collaboration with ongoing science rather than provision of general stories about what science really is, is to stay abreast of the developing science, to be in a position to respond to the questions it gives rise to, and to amend central biological concepts as new scientific insight makes necessary. If we can do that we will, I think, be doing very well.

References

Barnes, B., and J. Dupré. *Genomes and What to Make of Them.* Chicago: Chicago University Press, 2008.

Dawkins, R. *The Selfish Gene.* Oxford: Oxford University Press, 1976.

Dennett, D. C. *Darwin's Dangerous Idea: Evolution and the Meanings of Life.* New York: Simon and Schuster, 1995.

Doolittle, W. F. "Some Thoughts on the Tree of Life." *The Harvey Lectures, 2003-2004.* 111-128, 2005.

Dupré, J. "Natural Kinds and Biological Taxa." *The Philosophical Review.* 90: 66-91, 1981.

Dupré, J. *The Disorder of Things: Metaphysical Foundations of the Disunity of Science.* Cambridge, MA.: Harvard University Press, 1993.

Dupré, J., and M. O'Malley. "Varieties of Living Things: Life At The Intersection of Lineage And Metabolism." *Philosophy and Theory in Biology.* vol. 1, 2009. http://hdl.handle.net/2027/spo.6959004.0001.003

Kitcher, P. *Vaulting Ambition: Sociobiology and the Quest for Human Nature.* Cambridge, Mass: MIT Press, 1985.

Ladyman, J., and D. Ross. *Every Thing Must Go: Metaphysics Naturalized.* New York: Oxford University Press, 2007.

Odling Smee, J., K. Laland, and M. Feldman. *Niche Construction: The Neglected Process in Evolution.* Princeton: Princeton University Press, 2003.

O'Malley, M., and J. Dupré. "Fundamental Issues in Systems Biology." *BioEssays.* 27: 1270-1276, 2005.

O'Malley, M., and J. Dupré. "Size Doesn't Matter: Towards a More Inclusive Philosophy of Biology." *Biology and Philosophy.* 22: 155-191, 2007.

Wilson, E. O. *Sociobiology: The New Synthesis.* Cambridge, Ma.: Harvard University Press, 1975.

Selected Bibliography

Books

The Latest on the Best: Essays on Evolution and Optimality. Editor. Bradford Books/MIT Press, 1987.

The Disorder of Things: Metaphysical Foundations of the Disunity of Science. Harvard University Press, 1993.

Humans and Other Animals. Oxford University Press, 2002.

Darwin's Legacy: What Evolution Means Today. Oxford University Press, 2003; German translation, Suhrkamp Verlag, 2005; Spanish translation, Katz Editores, Buenos Aires, 2006.

Value-Free Science? Ideals and Illusions. Co-edited with H. Kincaid and A. Wylie. New York: Oxford University Press, 2007.

The Constituents of Life. (The Spinoza Lectures, University of Amsterdam). Amsterdam: Van Gorcum, 2008.

Genomes and What to Make of Them. (with Barry Barnes) Chicago University Press, 2008.

Selected Articles

"In Defence of Classification." *Studies in The History and Philosophy of the Biological and Biomedical Sciences.* 32: 203-219, 2001.

"On Human Nature." *Human Affairs.* 13: 109-122, 2003. Slovakian translation forthcoming for the inaugural issue of the *Magazine for Human Sciences.*

"The Miracle of Monism." In D. MacArthur and M. de Caro (eds.), *Naturalism in Question,* 36-58. Harvard University Press, 2004. Italian translation, Laterza, Roma, 2005.

"Understanding Contemporary Genomics." *Perspectives on Science.* 12: 320-338, 2004.

"Human Kinds and Biological Kinds: Some Similarities and Differences." *Philosophy of Science.* supp. vol.: 892-900, 2004.

"What's the Fuss about Social Constructivism?" *Episteme.* 1: 73-85, 2004.

"Are There Genes?" In A. O'Hear (ed.), *Philosophy, Biology and Life* (Royal Institute of Philosophy Supplements), 193-210. Cambridge: Cambridge University Press, 2005.

"Size Doesn't Matter: Towards a More Inclusive Philosophy of Biology." (with M. O'Malley) *Biology and Philosophy.* 22: 155-191, 2007.

"Metagenomics and Biological Ontology." (with M. O'Malley) *Studies in the History and Philosophy of the Biological and Biomedical Sciences.* 38: 834-846, 2007.

"What Genes Are, and Why There Are No 'Genes For Race.'" In B. A. Koenig, S. S.-J. Lee, and S. Richardson (eds.), *Revisiting Race in a Genomic Age.* Rutgers University Press, 2008.

3

Arthur Fine

Professor of Philosophy
Adjunct Professor of Physics and of History
University of Washington, USA

---

1. How were you initially drawn to philosophical issues regarding science?

I discovered the Socratic dialogues as an adolescent. At about the same time I learned the calculus, and also learned to build primitive radios. Philosophy and science have been on my mind ever since. My undergraduate training was in mathematics, with physics, operations research and philosophy as minor subjects. One of my mathematics advisors was Irving Segal, then in the middle of his analytical work on C*-algebras, work motivated in part by a conceptual analysis of what he regarded as essential in the quantum theory. Segal challenged me about my studies and life plans, emphasizing the importance of being serious about both. Later I studied the mathematical foundations of quantum mechanics with Karl Menger, a sometime member of the Vienna Circle. Menger was something of a provocateur, highlighting the conceptual problems of the quantum theory and always referring to quantum theorists as the "metaphysicians" of our time. He provoked me to serious philosophical reflections on science. I was fortunate in being able to pursue these reflections in my doctoral thesis, on quantum measurement and directed by Henry Mehlberg, himself a product of the famous Polish school in the methodology of science.

2. What, in your view, are the most interesting, important, or pressing problems in contemporary philosophy of science?

That's easy; normativity and values are where the most difficult and most important work needs to be done. Although neopositivism was rooted in certain normative concerns, emphasizing cognitive values and promoting science as a progressive social institution, it left a legacy of programs that invited thin formal analyses at levels of abstraction so high as to create dizziness rather than enlightenment (our familiar "theories" of confirmation, explanation, scientific laws, theory construction, etc.). The Vienna Circle's encounter with Wittgenstein marked a point at which many in the movement turned from dealing with the complexities of scientific theorizing and experimental practice to constructing (mostly) schematic treatments of the language of science. Twenty years later, cold war politics inclined others to shy away from thick issues, including science policy, where questions of values could not be avoided. Under the influence of Stephen Toulmin and Thomas Kuhn philosophy of science began to direct our attention to scientific communities bound together by the values inherent in ideals of natural order (Toulmin) or paradigms (Kuhn). More recently, prompted by constructivist and feminist critiques, and prodded by conservative over-reactions, mainstream philosophy of science has begun to examine the ways in which scientific practice, at all stages, involves norms and values.

3. How has your work offered original contributions to discussion on science? What does your work reveal that others fail to appreciate?

My work treats technical issues that arise concretely in scientific practice (measurement, locality, decoherence, probabilistic modeling, metric geometry, among others), and also general issues (NOA) related to the traditional problems of philosophy and philosophy of science (realism, relativism, objectivity and the like) often in historical context (e.g., Einstein on EPR). These agendas are connected by my belief that important aspects of science can profitably be seen as engaged with philosophy, and vice versa. Thus I reject a "two cultures" picture, with its neat division of labor into humanistic *versus* scientific. Similarly I think that two levels approaches to understanding science (external *versus* internal, or transcendental/constitutive *versus* empirical, or simply philosophical *versus* scientific) inevitably distort what goes on. I think of science as a reflective enterprise that integrates recognizably philosophical concerns with technical ones. But I reject Quinean naturalism, according to which, when science has had its say, there will be nothing left over for philosophy—or at any rate nothing very important. My rule of thumb is simply that good philosophy of science always connects with ongoing scientific practice. This is hardly original and today, increasingly, we see others pursuing similar agendas.

4. What is the relation between philosophy of science and scientific practice, science policy, or efforts for social justice? Can there be a more productive relation? Is this desirable?

Philosophy of science is dedicated to understanding scientific practice critically, and to promoting good practice. And scientific practice, like all fields of human endeavor, is normatively framed. All the sciences operate with frameworks of principles, standards, ideals, and values, which change and develop over time. They are always works in progress since, as John Dewey taught us, we learn in science how better to pursue it. These normative frameworks are not isolated. They are embedded in our general culture and its steady revaluations. Inevitably science policy, science education and questions of social justice become entangled with ongoing science. To mention just a few areas where values conflict, think of stem cell research, the continuing clash over teaching creationism (and its cousins), or the politics of climate change. A critical understanding of normative frameworks in scientific practice ought to help promote good practice both in research science and in public science.

5. Where do you see the field of philosophy of science to be headed? What are the prospects for progress regarding the issues you take to be most important?

I think the prominence of value issues will affect philosophy of science, aligning it more closely with value theory in a way that will profit both. Also, new areas in science—e.g., simulation science, nanotechnology, neuroscience—will enlarge the field of practice and yield fresh perspectives on how the sciences develop. Witness recent philosophical treatments of scientific fictions and modeling. These new areas also give rise to fresh interconnections with values and policy issues. In my own domain, philosophy of physics, I look forward to developments in quantum gravity: first to a loosening of the stranglehold of string theory, and then to the development of a philosophically important new synthesis. Overall, I think we will continue to move away from grand generalizations about science and formal analyses carried out in a thin atmosphere of abstraction, and move toward more situated ways of understanding scientific endeavors. As this movement proceeds, informative work in philosophy of science will need to balance sufficient generality with a decent range of particularity. Whatever the balance, our work intermingles description with normativity in ways that challenge the traditional fact–value distinction, or the distinction between a context of discovery and a context of justification. Thus we have a natural alliance with the several other disciplines who study science (anthropology, economics, education, history, sociology, rhetoric, and so on). As our understanding of normativity grows and percolates through the field I expect to see important changes, and lots of good work!

Selected Bibliography

Books

Bohmian Mechanics and Quantum Theory: An Appraisal. Co-edited with J. Cushing and S. Goldstein. Dordrecht: Kluwer, 1996.

The Shaky Game: Einstein, realism and The Quantum Theory. Chicago: University of Chicago Press, 1986; Revised Paperback Edition, 1988; Japanese translation, 1992; Second Edition, 1996.

PSA: 1990, Vols. 1 and 2. Co-edited with M. Forbes and L. Wessels. E. Lansing, MI: Philosophy of Science Association, 1990 and 1991.

PSA: 1988, Vols. 1 and 2. Co-edited with M. Forbes & J. Leplin. E. Lansing, MI: Philosophy of Science Association, 1988 and 1989.

PSA: 1986, Vols. 1 and 2. Co-edited with P. Machamer. E. Lansing, MI: Philosophy of Science Association, 1986 and 1987.

Selected Articles

"Bohr's Response to EPR: Criticism and Defense." *Iyyun, The Jerusalem Philosophical Quarterly.* 56: 31-56, 2007.

"Decoherence and the Foundations of Quantum Mechanics." (with M. Schlosshauer) In J. Evans and A. Thorndike (eds.), *Quantum Mechanics at the Crossroads*, pp. 125-48. New York: Springer Publishing Co., 2007.

"Relativism, Pragmatism and the Practice of Science." In C. Misak (ed.), *The New Pragmatists*, pp. 50-67. Oxford: Oxford University Press, 2007.

"Noncommuting Observables and Local realism." (with J. D, Malley) *Physics Letters A*, 347/1-3: 51-55, 2005. Also online at http://arxiv.org/abs/quant-ph/0505016.

"Realism, Beyond Miracles." (with A. Mueller) In Y. Ben Menahim (ed.), *Contemporary Philosophy in Focus: Hilary Putnam*, pp. 83-124. Cambridge:Cambridge University Press, 2004.

"The Einstein-Podolsky-Rosen Argument in Quantum Theory." In E. N. Zalta (ed.) *The Stanford Encyclopedia of Philosophy*, 2009, http://plato.stanford.edu/entries/qt-epr/

"Quantum Life: Interaction, Entanglement, and Separation." (with E. Winsberg), *Journal of Philosophy*. C: 80-97, 2003.

"The Viewpoint of No-one in Particular." *Proceedings and Addresses of The American Philosophical Association*. 72: 9-20, 1998; Reprinted as "Der Blickpunkt von niemand im besonderen." In M. Sandbothe (ed.), *Die Renaissance des Pragmatismus: Aktuelle Verflechtungen zwischen analytischer und kontinentaler Philosophie*, pp. 59-77. Weilerswist: Velbrück Wissenschaft, 2000; and in M. Sandbothe and W. Egginton (eds.), *The Pragmatic Turn in Philosophy*, pp. 115-129. Albany, NY: SUNY Press, 2004.

"A Local Hidden Variable Theory for the GHZ Experiment." (with L. Szabo) *Physics Letters A*. 295: 229-40, 2002. (Erratum, vol. 302: 345, 2002.) Also online at arXiv quant-ph/0007102

"Gauge Theory, Anomalies and Global Geometry: The Interplay of Physics and Mathematics." (with D. Fine)*Studies in History and Philosophy of Modern Physics*. 28: 1-18, 1997.

"Science Made Up: Constructivist sociology of Scientific Knowledge." In P. Galison and D. Stump (eds.), *The Disunity of Science: Boundaries, Contexts, and Power*, pp. 231-54. Stanford: Stanford University Press, 1996.

"The Scientific Image Twenty Years Later." *Philosophical Studies*. 106: 107-22, 2001.

"Indeterminism and the Freedom of the Will." In J. Earman et al. (eds.), *Philosophical Problems of the Internal and External Worlds*, pp. 551-72. Pittsburgh: University of Pittsburgh Press, 1993.

"Fictionalism." *Midwest Studies in Philosophy*. 18: 1-18, 1993; Reprinted in M. Suárez (ed.), *Fictions in Science: Philosophical Essays on Modeling and Idealization*. London: Routledge, forthcoming.

"Measurement and Quantum Silence." In S. French and H. Kamminga (eds.), *Correspondence, Invariance and Heuristics*, pp. 279-94. Dordrecht: Kluwer, 1993.

"Models, Chaos and Goodness of Fit." (with S. Kellert and M. Stone) *Philosophical Topics.* 18: 85-105, 1991.

"Piecemeal Realism." *Philosophical Studies.* 61: 79-96, 1991.

"Causes of Variation: Disentangling Nature and Nurture." *Midwest Studies in Philosophy.* 15: 94-113, 1990.

"Do Correlations Need To Be Explained?" In J. Cushing and E. McMullin (eds.), *Philosophical Consequences of Quantum Theory.* Notre Dame: U. of Notre Dame Press, 1989, pp. 175-94.

Plus over one hundred other articles and reviews in philosophical journals (including *Ethics, Journal of Philosophy, Nous, Mind, Philosophical Review, Philosophical Studies, Philosophy of Science, Synthése*), scientific journals (including *The American Journal of Physics, The Behavioral and Brain Sciences, Foundations of Physics, Journal of Mathematical Physics, Nature, Physical Review, Physics Letters A, Proceedings of The American Mathematical Society, Science*), and numerous anthologies.

4
Allan Franklin

Professor of Physics
University of Colorado, USA

---

1. How were you initially drawn to philosophical issues regarding science?

I began my career as an experimental particle physicist and later changed my research field to the history and philosophy of science, with a particular emphasis on experiment in physics. This is not a typical career path and it began while I was an undergraduate at Columbia College. In 1957, T.D. Lee, then a professor at Columbia, and C.N. Yang, then at the Institute for Advanced Study, were awarded the Nobel Prize in Physics for their suggestion that parity, or mirror or space-reflection symmetry was not conserved in the weak interactions. They had also suggested several experimental tests of their hypothesis, whose successful completion (Wu, Ambler et al., 1957; Garwin, Lederman et al., 1957; Friedman and Telegdi, 1957), led to both the discovery of parity nonconservation and to the prize.[1] This was regarded as such a significant achievement that physics classes at Columbia were stopped and our instructors explained the details to us. This made a considerable impression on me and its significance for my subsequent career would appear only later. In addition, I decided that as a physics major I should have some understanding of the philosophy underlying my discipline and managed to get permission to take a graduate course in the philosophy of science with Phillip Frank. One point that I remember from that course was Frank's view that the role of the philosopher of science was to mediate between philosophy and science—to explain philosophy to the scientists and science to the philosophers. I pursued this interest by taking a second course in philosophy of science, with

---

[1] For a detailed history of this episode see Franklin (1979).

Douglas Gasking, while I was a graduate student in physics at Cornell University. At the time I intended to be an experimental particle physicist, and philosophy of science was only a minor interest.

After obtaining my degree and after a stint as a postdoctoral research associate at Princeton, I became an assistant professor of physics at the University of Colorado. It was a time, the late 1960s and early 1970s, when things were more flexible and one could pursue other interests. Along with my friend and colleague, Howard Smokler, a philosopher of science, I taught a course on the similarities and differences between the episodes involving Galileo and J. Robert Oppenheimer. As part of my preparation for that course, I did some reading on the history of medieval science in order to understand the background to Copernicus and Galileo. Having been trained as a physicist, I had thought that the Middle Ages were indeed a dark age for science. I was fascinated and intrigued to discover that there was something that I could regard only as high quality science during that period. I subsequently obtained a position as visiting professor at the New School of Liberal Arts, City University of New York, where everything was taught by historical period. My job was to teach the history of medieval science.[2] During that year I got an idea for what I thought would be a pedagogical article for my physics colleagues on how the principle of inertia developed during the medieval period. The project got out of hand and became a monograph, *The Principle of Inertia in the Middle Ages* (1976).

It quickly became apparent to me that I did not have either the training or background to be a medievalist, although I did take an introductory Latin course, but that what I did know was something about experiments in physics. At this time the enormous impression made on me by the discovery of parity nonconservation and of the role of "crucial" experiments in that episode returned and I decided to look at the history of that episode. I had also read Thomas Kuhn's *The Structure of Scientific Revolutions* in which he doubted the existence of such crucial experiments. There had also been stories circulating within the physics community of

---

[2] I must acknowledge here my debt to Professor Edward Grant of Indiana University. I wasn't really prepared to teach such a course. With the encouragement of Joan Cadden, a former student of Ed's, I called Ed asking for assistance. Ed provided both bibliographies and reprints of some of his own work which were invaluable for both my teaching and subsequent work. He later suggested that I write a paper on impetus theory (Franklin, 1977).

earlier experiments, which had also shown parity nonconservation, but whose significance was not perceived at the time. I was intrigued by the idea that an experiment could be crucial at one time and relatively unimportant at another. It was the investigation of these issues that led to my first work on the history and philosophy of experiment, "The Discovery and Nondiscovery of Parity Nonconservation" (1979). As they say, the rest is history.

I cannot close this section without acknowledging my debt of gratitude to the Department of History and Philosophy of Science, Chelsea College, University of London and to its head, Professor Heinz Post. In the spring of 1978 I spent my sabbatical leave as a visiting researcher in the department.[3] I arrived literally carrying my parity conservation paper in my backpack. Heinz made it quite clear to me that in return for my desk and library privileges my only responsibility was to attend and participate in the weekly Thursday department seminars. I learned an enormous amount of philosophy of science in the give and take, some might call it an intellectual free-for-all, of those seminars. Several of us from Chelsea were also regular participants in the Tuesday seminars at the London School of Economics. It was there that I met Colin Howson, with whom I have enjoyed both a long friendship and a fruitful research collaboration. Whatever I know of Bayesianism I learned in numerous walks across Hampstead Heath with Colin. It was truly peripatetic philosophy.

2. What, in your view, are the most interesting, important, or pressing problems in contemporary philosophy of science?

Although I do not feel qualified to name the most important problem in all of philosophy of science, I do have some thoughts on an important problem in the area in which I work, namely, the history and philosophy of experiment. I believe that the question how experimental evidence can provide legitimate grounds for belief in our theories of nature is not only an important question for philosophers of science, but also for the general public. As I will discuss later, I believe that philosophers of science have not done enough to inform the public about the relevance of evidence to theory and this has led to unfortunate consequences for our society.

---

[3] Noretta Koertge of Indiana University was instrumental in arranging for my initial position.

4. Allan Franklin

I will begin, however, with some of the issues for philosophers of science. Many practitioners have adopted the Bayesian approach to scientific reasoning. (For a detailed discussion of this approach see Howard and Urbach, 1993.) Briefly stated, Bayesianism is the view that we have varying degrees of belief in hypotheses and that these beliefs obey the probability calculus. Richard Cox has shown that under quite plausible and reasonable assumptions this is, in fact, correct.[4] If this is so then our degrees of belief follow Bayes's Theorem[5]

$$P(h|e) = \frac{P(e|h)P(h)}{P(e)} \text{ where}$$
$$P(e) = P(e|h)P(h) + P(e|\neg h)P(\neg h).$$

$P(h|e)$ is the probability of h given e, which is called the posterior probability, whereas $P(h)$ and $P(e)$ are the prior probabilities of h and e, respectively, and $\neg h$ is "not h."

Because $P(e)$ is always less than one this leads to the result that when h entails e, $(P(e|h) = 1)$, $P(h|e)$ is always greater than the $P(h)$. Thus, we have the intuitively satisfying conclusion that the observation of evidence entailed by a hypothesis should strengthen your belief in that hypothesis. This would seem to be the essence of rationality. If observation of evidence entailed by a hypothesis doesn't strengthen your belief then what should? Other successes of the Bayesian view are that it gives the conclusion that a variety of experimental results offers more support for a hypothesis than the evidence provided by the repetition of the same experiment,[6] that adding an additional statement to a hypothesis reduces the

---

[4] Cox argues that probable inference satisfies the probability calculus if: 1) The degree of belief in an inference on given evidence determines the degree of belief of its contradictory on the same evidence; 2) The degree of belief on given evidence that both of two inferences are true is determined by their separate degrees of belief, one on the given evidence, the other on this evidence with the additional assumption that the first inference is true; and 3) The degrees of belief satisfy the logical rules of Boolean algebra. One may also reasonably substitute "degree of belief" for "probable inference." Cox is discussing a comment by John Venn, who in discussing "quantities of belief," asked whether the laws according to which the belief is produced and varied are the same. Cox answered yes, and argued that probability provided "one harmonious scheme" (Cox, 1961).

[5] Bayes's Theorem is a theorem of the probability calculus and is true independent of whether one adopts the view that we have degrees of belief that satisfy that calculus.

[6] A homey example is as follows. If both our watches are running and we

support given by the evidence to the original hypothesis alone (the tacking paradox), and it provides a way of estimating the importance of different pieces of evidence.[7]

Several objections have been raised to Bayesianism. They include the problem that, as of yet, no one has suggested an acceptable method of objectively establishing the prior probabilities required. The best we have been able to do is to assign subjective probabilities as to the fair odds of a hypothesis. Critics object that this introduces an unacceptable subjectivity into science. I note, however, that one does not reject deductive logic because it has no way of determining the truth of its premises. I also suggest that in the actual practice of science if one has initial probability distributions that do not assign zero probability to plausible alternatives or probability one to a single hypothesis then the initial priors will not, after a reasonable amount of evidence has been acquired, determine which the best supported hypothesis is. This is not, of course, a proof. Other criticisms include the problem of "old evidence," if evidence is already known it cannot provide support for a hypothesis because P(e) is one; and the problem of convergence of belief as evidence is acquired.[8]

Recently, Deborah Mayo (1996) has both criticized Bayesianism and offered her own alternative, the error-statistics approach. In Mayo's view a hypothesis is confirmed only if it has passed a severe test. For Mayo this means that the probability of the

---

want to know the correct time, it is better if we compare watches than if I look at my watch twice.

[7] If an experimental result is either very improbable, or very improbable on any hypothesis except the one under consideration, then observation of that result provides substantial support for the hypothesis. Richard Cox presents a very nice example of this from Shakespeare's *Macbeth*.

"Consider Macbeth's reasoning about the witches who hailed him on the desolate heath as thane of Glamis and Cawdor and thereafter king. At first he was incredulous and said

'By Sinel's death I know that I am thane of Glamis;
But how of Cawdor? The thane of Cawdor lives,
A prosperous gentleman; and to be king
Stands not within the prospect of belief,
No more than to be Cawdor.'

Farther along the way he met King Duncan's messengers and learned that he had in truth become thane of Cawdor. So he was persuaded that the witches knew what they were talking about and the more so because the prediction just confirmed had been so improbable before" (Cox, 1961, p. 92).

[8] For a full discussion of the objections to Bayesianism as well as Bayesian answers to those objections see Howson and Urbach (1993).

evidence given the hypothesis is much larger than the probability of the evidence if the hypothesis is false, $P(e|h) >> P(e|\neg h)$. Mayo claims that this avoids the problem of assigning prior probabilities. In point of fact, it does not. We may write $P(e|\neg h) = [1/P(\neg h)] \sum_i P(e|h_i) P(h_i)$, where the h_i are the alternatives to h.

Thus, Mayo cannot avoid using prior probabilities.[9] Mayo admits that using her condition for support and Bayes's Theorem guarantees that $P(h|e) > P(h)$, so that Bayesians will also conclude that the hypothesis is supported. In addition, Mayo requires that the probabilities be objective.[10]

Although Mayo's view has been adopted by many philosophers of science, I believe it has several severe problems. Perhaps the most serious is that Mayo commits what is known as the "base-rate fallacy." By neglecting the prior probability of the hypothesis she arrives at what can only be called bizarre conclusions. Consider the following example. Suppose one has a test for a disease which always gives a positive result if you have the disease, $(P(e|h) = 1)$. Suppose also that the test also gives a positive result five percent of the time if you don't have the disease, $P(e|\neg h) = 0.05$. This certainly satisfies Mayo's criterion, $P(e|h) >> P(e|\neg h)$, so that if you get a positive result the hypothesis that you have the disease is confirmed. Should you undergo treatment? The answer is that you really don't know. Without considering the prior probability that you have the disease you have no way of making a reasonable judgment. Suppose that, in fact, the disease affects only one person in a thousand, $(P(h) = 0.001)$. A simple calculation using Bayes's Theorem yields a probability of two percent that you have the disease given that you had a positive test result. A simple example will make this clear. Suppose we have a sample of a thousand people and you administer the test to each of them and send those that have a positive result into a room. On average, one person of the thousand will have the disease and have a positive result. However, fifty of the remaining 999 will also have a positive result but won't have the disease. Your room will contain 51 people, only one of whom has the disease.[11]

---

[9] I am grateful to Colin Howson for pointing this out to me.

[10] For Mayo this requires that the probabilities be frequencies.

[11] I am grateful to Michel Janssen for providing this example during a very heated debate late at night during a Seven Pines Symposium.

In addition to rehearsing the standard objections to Bayesianism, Mayo emphasizes two related criticisms: 1) that Bayesians are unable to provide a method for obtaining an objective result for $P(e|\neg h)$, whereas her error-statistics approach yields a definite result and 2) that the error-statistics approach is used extensively in the practice of science whereas Bayesianism is not.[12] Mayo presents an example from high-energy elementary particle physics in which the only alternative considered was that the result was due to a chance fluctuation in the experimental background.[13] This is, in fact, the only type of case in which the error-statisticians can produce a definite result and avoid the use of prior probabilities.[14] However, the same calculation is also available to a Bayesian, and there is no advantage to the error statistician. More often, however, a chance fluctuation in background is not the only alternative, but other physical processes are discussed which might mimic the observed effect. As noted above, one must now invoke prior probabilities as well as estimates of $P(e|\neg h)$. Consider the experiment of Christenson et al. (1964), which measured the decay $K_2^0 \to \pi^+\pi^-$, and gave evidence for CP violation (Combined space-reflection and particle-antiparticle symmetry). The experimenters considered the possibility that their result was caused by a fluctuation in the background, but they also considered other physical processes that might mimic the decay. One possibility was $K_{\mu 3}$ or K_{e3} decay, in which the energy distribution of the neutrino emitted in such decays was very narrow. They noted that no plausible physical mechanism existed for such a process. A second alternative was $K_L^0 \to \pi^+\pi^-\gamma$ decay in which the energy of the γ ray was less than 1 MeV, whereas 209 MeV was available. Once again the experimenters noted that they knew of no physical process that could do this. In both cases the experimenters concluded quite reasonably that these processes were very unlikely to produce their result. The probabilities could only be estimated, not calculated, and here Bayesianism has a considerable advantage. In fact, because there are no objective

---

[12] I note here the increasing use of Bayesian statistical methods in science.

[13] Other examples exist. In the recent experiment by the Sudbury Neutrino Observatory (SNO) group on neutrino oscillations, they state that the probability that a downward fluctuation of the Super-Kamiokonde result would produce the SNO result is 0.04 %.

[14] The error statistician must also argue that no other alternatives had any significant probability. This would seem to require the use of prior probabilities.

probabilities, Mayo and the error statisticians cannot deal with this type of case.

I believe that the problem of how experimental evidence can provide support for our theories and hypotheses is central to philosophy of science. I have presented the view I favor, namely Bayesianism, as well as its most popular recent criticism. The issue is still unresolved. As I remarked in Footnote 11, the issue can still arouse heated debate.

3. How has your work offered original contributions to discussion on science? What does your work reveal that others fail to appreciate?

I believe that my most significant contribution to the discussion of science has been my work on the epistemology of experiment, a set of strategies that provide us with good reasons for belief in the correctness of experimental results. I believe that science is a reasonable enterprise based on experimental evidence, criticism, and rational discussion. It provides us with knowledge of the physical world and it is experiment that provides the evidence which grounds that knowledge. Experiment plays many roles in science. One of its important roles is to test theories and to provide the basis for scientific knowledge. It can also call for a new theory, either by showing that an accepted theory is incorrect, or by exhibiting a new phenomenon which needs explanation. Experiment can provide hints toward the structure or mathematical form of a theory and it can provide evidence for the existence of the entities involved in our theories. It can also measure quantities that theory tells us are important. Finally, it may also have a life of its own, independent of theory. Scientists may investigate a phenomenon just because it looks interesting. This will also provide evidence for a future theory to explain.

If experiment is to play these important roles in science then we must have good reasons to believe experimental results. I present below an epistemology of experiment, a set of strategies that provides reasonable belief in experimental results. Scientific knowledge can then be reasonably based on these experimental results.

It has been more than two decades since Ian Hacking asked, "Do we see through a microscope?" (Hacking 1981). Hacking's question really asked how do we come to believe in an experimental result obtained with a complex experimental apparatus? How do we distinguish between a valid result and an artifact created by that apparatus? Hacking provided an extended answer in the second half of *Representing and Intervening* ..(1983). He pointed out that observations remain robust despite changes in the theory of the apparatus or in the theory of the phenomenon. His illustration was the continuous belief in microscope images despite the major change in the theory of the microscope when Abbe pointed out the importance of diffraction in its operation. One reason Hacking gave for this is that in making such observations the experimenters intervened. They manipulated the object under observation. Thus, in looking at a cell through a microscope one might inject fluid into the cell or stain the specimen. One expects the cell to change shape or color when this is done. Observing the predicted effect strengthens our belief in both the proper operation of the microscope and in the observation. This is true in general. Observing the predicted effect of an intervention strengthens our belief in both the proper operation of the experimental apparatus and in the observations made with it.

Hacking also discussed the strengthening of one's belief in an observation by independent confirmation. The fact that the same pattern of dots, dense bodies in cells, is seen with "different" microscopes, i.e. ordinary, polarizing, phase-contrast, fluorescence, interference, electron, acoustic etc., argues for the validity of the observation. Hacking correctly argues that it would be a preposterous coincidence if the same pattern of dots were produced in two totally different kinds of physical systems. Different apparatuses have different backgrounds and systematic errors, making the coincidence, if it is an artifact, most unlikely. If it is a correct result, and the instruments are working properly, the agreement of results is understandable.

Hacking's answer is correct as far as it goes. It is, however, incomplete. What happens when one can perform the experiment with only one type of apparatus, such as an electron microscope or a radio telescope, or when intervention is either impossible or extremely difficult? Other strategies are needed to validate the observation. These may include:

1) Experimental checks and calibration, in which the experimental apparatus reproduces known phenomena. For example, if we wished to argue that the spectrum of a substance obtained with a new type of spectrometer is correct, we might check that this new spectrometer could reproduce the known Balmer Series in hydrogen. If we correctly observe the Balmer Series then we strengthen our belief that the spectrometer is working properly.

This also strengthens our belief in the results obtained with that spectrometer. If the check fails then we have good reason to question the results obtained with that apparatus.

2) Reproducing artifacts that are known in advance to be present. An example of this comes from experiments to measure the infrared spectra of organic molecules (Randall, Fowler et al. 1949). It was not always possible to prepare a pure sample of such material. Sometimes one had to place the substance in an oil paste or in solution. In such cases, one expects to observe, superimposed on the spectrum of the substance, the spectrum of the oil or the solvent, which one can compare with the known spectrum of the oil or the solvent. Observation of this artifact gives confidence in other measurements made with the spectrometer.

3) Elimination of plausible sources of error and alternative explanations of the result (the Sherlock Holmes strategy).[15] Thus, when scientists claimed to have observed electric discharges in the rings of Saturn, they argued for their result by showing that it could not have been caused by defects in the telemetry, by interaction with the environment of Saturn, by lightning, or by dust. The only remaining explanation of their result was that it was due to electric discharges in the rings. There was no other plausible explanation of the observation. In addition, the same result was observed by both Voyager 1 and Voyager 2. This provided independent confirmation. Often, several epistemological strategies are used in the same experiment.

4) Using the results themselves to argue for their validity. Consider the problem of Galileo's telescopic observations of the moons of Jupiter. Although one might very well believe that his early telescope might have created spots of light, it would have been extremely implausible that the telescope would create them so that they would appear to be a small planetary system with eclipses and other consistent motions. It would have been even more implausible to believe that the created spots would satisfy Kepler's Third Law ($R^3/T^2=$ constant).[16] In this case one is arguing that there was no plausible malfunction of the apparatus, or background, which would explain the observations.

5) Using an independently well-corroborated theory of the phe-

---

[15] As Holmes remarked to Watson, "How often have I said to you that when you have eliminated the impossible, whatever remains, *however improbable, must be the truth*" (Conan Doyle, 1967).

[16] Kepler's Third Law was not available when Galileo made his observations, but it was an argument that could have been used later.

nomena to explain the results. This was illustrated in the discovery of the $W^{\pm}$, the charged intermediate vector boson required by the Weinberg-Salam unified theory of electroweak interactions. Although these experiments used very complex apparatuses and used other epistemological strategies (see Franklin, 1986, pp. 170-172, for details), I believe that the agreement of the observations with the theoretical predictions of the particle properties helped to validate the experimental results. In this case the particle candidates were observed in exactly the type of events that theory predicted. In addition, the measured particle mass found in the two experiments (note the independent confirmation), was in good agreement with the theoretical prediction. It was very improbable that any background effect, which might mimic the presence of the particle, would be in agreement with theory.

6) Using an apparatus based on a well-corroborated theory. In this case the support for the theory passes on to the apparatus based on that theory. This is the case with both the electron microscope and the radio telescope, whose proper operation is based on a well-supported theory, although other strategies may also be used to validate the observations.

7) Using statistical arguments. An interesting example of this arose in the 1960s when the search for new particles and resonances occupied a substantial fraction of the time and effort of those physicists working in experimental high-energy physics. The usual technique was to plot the number of events observed as a function of the invariant mass of the final-state particles and to look for bumps above a smooth background. The usual informal criterion for the presence of a new particle was that it resulted in a three standard-deviation effect above the background, a result that had a probability of 0.27% of occurring in a single bin. This criterion was later changed to four standard deviations, which had a probability of 0.0064% when it was pointed out that the number of graphs plotted each year by high-energy physicists made it rather probable, on statistical grounds, that a three standard-deviation effect would be observed.

8) Using "blind" analysis, a strategy for avoiding possible experimenter bias. This can include excluding the region of interest when setting the selection criteria that will be applied to the data. One can also add a random number to the measured parameter so that the value of the result does not bias the analysis of the data. (For details see Franklin, 2002, Chapter 6.)

These strategies along with Hacking's intervention and independent confirmation provide an epistemology of experiment.

Although all of the illustrations of the epistemology of experiment come from physics,[17] David Rudge (1998, 2001) and Marcel Weber (2005) have shown that similar strategies are also used in biology. Similar arguments have been made in applying the epistemology of experiment to computer simulation by Wendy Parker (2006) and by Eric Winsberg (1999; 2001; 2003; 2006).

4. What is the relation between philosophy of science and social issues?

As I indicated earlier, I believe that philosophy of science has important implications for several social issues which involve the relationship between theory and evidence. These include the evolution-creationism or evolution-intelligent design controversy, and the debate concerning global warming. I will restrict my comments to the former issue.

In the case of evolution versus creationism I believe that both philosophers of science and scientists need to do more to educate the public about both the general nature of science and more specifically about the scientific evidence. Several of them have already performed a valuable service in testifying about this issue in the recent Dover, Pennsylvania trial and in previous trials in Louisiana and Arkansas, but I believe that more is needed. Although the evolutionary side, supported by overwhelming evidence,[18] has won the court battles it is clear that we are losing the war. Reports in the popular press state that a significant majority of the population of the United States, including former President Bush, believes that both views should be taught in science classes, and forty percent believe that creationism or intelligent design should be taught alone.

In discussions about this issue, one often hears the statement "Evolution is only a theory." It is true that evolution is a theory, but to appropriate George Orwell, "All theories are equal, but some theories are more equal than others." This is because those theories have overwhelming evidential support, as is the case with evolutionary theory. It is important for those of us who either engage in science or who study it, to work to change the public's view, at all levels of education, not only at the university level,

---

[17] For a very nice example from physics, the discovery of the top quark, see Staley (2004).
[18] See for example Miller (1999) and Fairbanks (2007).

our usual stamping grounds.

I am not sure that this is an issue that can, or will, be decided by experimental evidence and reasoned and critical discussion, but I believe we must try harder. Perhaps we need to be less tolerant. When someone says that, in fairness, we should teach the controversy, we must make it clear that there is no scientific controversy and that the evidence overwhelmingly supports one side and not the other. As Kenneth Miller (1999) has eloquently argued, this need not have an adverse effect on one's religious beliefs.

5. Where is the philosophy of science headed?

One of the pleasant aspects of contemporary philosophy of science has been its resistance to trendiness and postmodernism, unlike other aspects of science studies. It has continued to work on its traditional important problems such as foundational issues in the sciences and questions such as epistemology and the relationship between evidence and hypothesis. As I mentioned earlier these issues have been recently extended beyond physics and biology to a welcome burgeoning literature on the philosophy of chemistry. It has also been extended other sciences, including the social sciences.[19] I expect these trends to continue.

References

Christenson, J. H., J. W. Cronin, V. L. Fitch, et al. "Evidence for the 2π Decay of the K_2^0 Meson." *Physical Review Letters*. 13: 138-140, 1964.

Conan Doyle, A. "The Sign of Four." In W. S. Baring-Gould (ed.), *The Annotated Sherlock Holmes*. New York: Clarkson N. Potter, 1967.

Cox, R. T. *The Algebra of Probable Inference*. Baltimore: The Johns Hopkins Press, 1961.

Fairbanks, D. J. *Relics of Eden: The Powerful Evidence of Evolution in Human DNA*. Amherst, N.Y.: Prometheus Boooks, 2007.

Friedman, J. L., and V. L. Telegdi. (1957). "Nuclear Emulsion Evidence for Parity Nonconservation in the Decay Chain pi - mu - e." *Physical Review*. 105: 1681-1682, 1957.

---

[19] The recently published *Handbook of the Philosophy of Science* (Gabbay et al. 2007) includes volumes on these subjects.

Gabbay, D. M., P. Thagard, and J. Woods. (eds.) *Handbook of the Philosophy of Science*. Amsterdam: Elsevier, 2007.

Garwin, R. L., L. M. Lederman, and M. Weinrich. "Observation of the Failure of Conservation of Parity and Charge Conjugation in Meson Decays: The Magnetic Moment of the Free Muon." *Physical Review.* 105: 1415-1417, 1957.

Hacking, I. *Representing and Intervening.* Cambridge: Cambridge University Press, 1983.

Howson, C., and P. Urbach. *Scientific Reasoning: The Bayesian Approach.* Chicago: Open Court, 1993.

Kuhn, T. *The Structure of Scientific Revolutions.* University of Chicago Press, 1962.

Mayo, D. *Error and the Growth of Experimental Knowledge.* Chicago: University of Chicago Press, 1996.

Miller, K. R. *Finding Darwin's God.* New York: HarperCollins, 1999.

Parker, W. "Understanding Pluralism in Climate Modeling." *Foundations of Science.* 11: 349-368, 2006.

Rudge, D. W. "Kettlewell from an Error Statistician's Point of View." *Perspectives on Science.* 9: 59-77, 2001.

Weber, M. *Philosophy of Experimental Biology.* Cambridge: Cambridge University Press, 2005.

Winsberg, E. "Sanctioning Models: The Epistemology of Simulation." *Science in Context.* 12: 275-93, 1999.

Winsberg, E. "Simulations, Models, and Theories: Complex Physical Systems and their Representations." *Philosophy of Science.* 68: S442-S454, 2001.

Winsberg, E. "Simulated Experiments: Methodology for a Virtual World." *Philosophy of Science.* 70: 105-125, 2003.

Winsberg, E. "Models of Success vs. the Success of Models: Reliability Without Truth." *Synthese.* 152: 1-19, 2006.

Wu, C. S., E. Ambler, R. W. Hayward, et al. "Experimental Test of Parity Nonconservation in Beta Decay." *Physical Review.* 105: 1413-1415, 1957.

Selected Bibliography

Books

The Principle of Inertia in the Middle Ages. Boulder: Colorado Associated University Press, 1976.

The Neglect of Experiment. Cambridge: Cambridge University Press, 1986.

Experiment, Right or Wrong. Cambridge: Cambridge University Press, 1990.

The Rise and Fall of the Fifth Force. New York: American Institute of Physics, 1993.

Can That Be Right? Essays on Experiment, Evidence, and Science. Dordrecht: Kluwer Academic Publishers, 1999.

Are There Really Neutrinos? An Evidential History. Cambridge, MA: Perseus Books, 2000.

Selectivity and Discord: Two Problems of Experiment. Pittsburgh: University of Pittsburgh Press, 2002.

No Easy Answers: Science and the Pursuit of Knowledge. Pittsburgh: University of Pittsburgh Press, 2005.

Wrong for the Right Reasons. Co-edited with J. Z. Buchwald. Archimedes. Dordrecht: Springer, 2005.

Ending the Mendel-Fisher Controversy. (with A. W. F. Edwards, D. J. Fairbanks, et al.) Pittsburgh: University of Pittsburgh Press, 2008.

Selected Articles

"The Discovery and Nondiscovery of Parity Nonconservation." *Studies in History and Philosophy of Science.* 10: 201-257, 1979.

"Why Do Scientists Prefer to Vary Their Experiments?" (with Colin Howson) *Studies in History and Philosophy of Science.* 15: 51-62, 1984.

"It Probably is a Valid Experimental Result: A Bayesian Approach to the Epistemology of Experiment." (with Colin Howson) *Studies in History and Philosophy of Science.* 19: 419-427, 1988.

"The Appearance and Disappearance of the 17-keV Neutrino. " *Reviews of Modern Physics.* 67: 457-490, 1995.

"Experiment in Physics." In Edwrad R. Zalta (ed.), *Stanford Encyclopedia of Philosophy*, 1999; Major revision, 2007. http://plato.stanford.edu.

"William Wilson and the Absorption of Beta Rays." *Physics in Perspective*. 4: 50-70, 2002.

"Doing Much About Nothing." *Archive for the History of Exact Sciences*. 58: 323-379, 2004.

"Introduction: Beyond Disunity and Historicism." (with J. Z. Buchwald) In J. Z. Buchwald and A. Franklin (eds.), *Wrong for the Right Reasons*, 1-16. Dordrecht: Springer, 2005.

"The Role of Experiments in the Natural Sciences: Examples from Physics and Biology." In T. A. F. Kuipers (ed.), *General Philosophy of Science: Focal Issues*. Amsterdam: Elsevier, 2007.

"Experiment in Physics." In Edwrad R. Zalta (ed.), *Stanford Encyclopedia of Philosophy*, 1999; Major revision, 2007. http://plato.stanford.edu.

5
Peter Galison

Joseph Pellegrino University Professor
Harvard University, USA

---

1. How were you initially drawn to philosophical issues regarding science?

It was Vietnam time—politics and philosophy elbowed into discussions all over the place. In high school I used to hang out among the bookstores up near Columbia University, poking around the philosophy, politics, and literature sections. Everyone was talking to everyone—students, professors, sidewalk philosophers. In the midst of all that political turbulence, someone told me to read Herbert Marcuse's *One-Dimensional Man* and from there to Nietzsche. One day one of the graduate students in philosophy, having heard I was interested in science, recommended that I read Willard van Quine's *From a Logical Point of View*. That was the first analytic philosophical book I picked up—and began, haltingly to read my way through, haltingly because I had very little idea what he was arguing against, a bit like hearing the punch line to a great joke, minus the windup. Someone else told me to read Hans Reichenbach's *Symbolic Logic*. That was easier to work through for me, but equally opaque insofar as I also had no idea what it was *against*.

It wasn't a terrific moment for science—the demonstrators were chanting against plastic shrapnel and napalm—me too—and I remember rather furtively meeting with my physics teacher after school to do some more advanced work. Somehow I'd gotten a few years ahead of myself, didn't feel ready to go to university, and spent a year in Paris. The chaos of course was even more intense there in 1972; I'd just turned seventeen, was working in a plasma physics laboratory at Ecole Polytechnique, and trying desperately to figure out what the hell was going on around me. Physics was the least problematic; I could sit down and work out the equations or fiddle with the Q-machine that was pumping ion waves through a plasma under the deft supervision of a kind young physicist, Dominique Gresillon. Philosophically, Paris was very much still in the long late 1960s, and the student world was awash in ideas, some quite interesting, others dogmatic, brittle, and gone as soon as they were formulated. There was a fabulous bookshop called Maspero I used to go to late at night: the political arguments never ended, this Trotskyist faction battling that one; one night everyone paraded over to a nearby auditorium to debate what Kafka had to say about the modern bureaucracy. It was Foucault one day and Althusser and Lacan the next; my young student philosophy friends were struggling to sort out what role the logos had—that is, when they weren't throwing bloodwurst across the dining room at the nearby preparatory school, Henri IV—and my head was still reeling from the demonstrations ("Une seule so-lu-tion... la ré-vo-lu-tion"... "Nixon assassin, Pompidou complice.") Rocks flew. I met an actual agent provocateur at a party one night—told me his favorite way to precipitate the attack of the dreaded head-clonking CRS riot police was to smash a shop window, so he always carried some bricks. Great, I thought, what a guy.

I sat in all-night cafes and read like I'd die if I stopped. I read 18th century novels and yesterday's political broadsheets. And stumbled across Thomas Kuhn's *Structure of Scientific Revolutions* in the summer of 1972 somehow; and then, after my year of Paris, suddenly was a second-year student at Harvard in the fall of 1973. I saw Gerald Holton's *Thematic Origin of Scientific Thought* in the window of a bookstore during those first days of September and gobbled it up—the essays on Einstein especially. Because I'd parachuted in I as a second-year student had to sort out what to major in within a few days and first chose, well, never mind, I ended up in the history of science and immediately began to look over at the philosophy department and happily plunked myself down in a class taught by Hilary Putnam. His essays on realism, reductionism, and physics brought me into philosophy as nothing else had—and began a conversation that has continued now for thirty-five years.

Whatever its faults were back in those years, history of science was never far from philosophy; the two disciplines eyed each others a little suspiciously, true enough, but people read. We argued about Imre Lakatos and Musgrave's *Criticism and the Growth of Knowledge* and from there on to the gamut of debates that still swirled around Lakatos's own work, around Popper, and around the myriad forms of realism and anti-realism. By the time I got to Cambridge in 1977-78 for their M.Phil. program I was completely immersed in a certain kind of philosophy—and delighted to study with Mary Hesse and Gerd Buchdahl.

With Mary Hesse, I worked on the philosophy of physics, specifically on some problems of the measurement problem (EPR) in concert with special relativity. I liked best, still do, the sparkling clarity of Bernard d'Espagnat's *Conceptual Foundations of Quantum Mechanics*. And with Buchdahl I took a course (there were two of us students) on Kant's *First Critique*. I loved the tutorial—but I had to make him promise not to explain Kant by reference to Heidegger—since this mainly left me twice confused. Not all was sweetness and light: In those days it was something like a crime to cross disciplines there and I once got hauled into the Master of Churchill's residence to explain myself as to why I had been seen, (someone had denounced me, Stasi style) taking a relativistic quantum course at the Cavendish. Luckily, Mary Hesse rescued me, explaining that I needed it for my thesis about the measurement problem in relativistic QM. The Authorities told me they'd let it go. But it better not happen again. So from then on I sat in the back of the big classroom with a hat. The Dirac equation as samizdat.

Before going to Cambridge, I had to sort out graduate school. The most obvious solution was to go to Princeton where Tom Kuhn was teaching. We talked a lot during the summer of 1977—he was living in a small place at the Centerfor Advanced Study in the Behavioral Sciences finishing his *Black-Body Theory* book, I was working on the Einstein Papers, then still in a rather parlous state, but coming into some kind of order under the supervision of John Stachel. With Kuhn we talked about Planck's combinatorics—I liked the manuscript of *Black-Body Theory* a lot—though I didn't, and still don't, much appreciate the way Kuhn presented Planck's entropy calculations. We talked about the relation of his work to Martin Klein's earlier studies on Planck's 1900 papers. Somehow in 1977-78, I got drafted into writing about all this for the *British Journal in the Philosophy of Science*—trying to sort out those arguments between Kuhn and Klein felt a bit as if I was auguring a huge hole in the bottom of my rowboat. In any case, Kuhn decided to leave Princeton for MIT, MIT was just building a program, and Yale's president, Hanna Gray, had just shut down its history of science. So I decided to come back to Harvard after Cambridge.

One day in an experiment course taught by Bob Pound, (one of the inventors of nuclear magnetic resonance), I was fiddling with the probe, trying to get a signal out of the NMR machine. The oscilloscope showed a spike and I thought "good"—that's what I was after, the resonance. Pound came over and told me it was spurious—and began adjusting the controls until a roughly similar spike appeared, but this one less bright, slightly trembling, more diffuse. He turned to me and said "that's it, that's the real signal." I stared for a long time at that scope—trying to figure out how he had known that that, rather than the spurious spike, had been the real McCoy.

Just about the same time, electroweak unified field theory—the Weinberg-Salam-Glashow account in particular—was also on the laboratory floor. I knew Weinberg and Glashow—they were hard to miss in the physics department. And I knew and had taken classes from some of the experimentalists who had been trying to sort out if the theory had experimental legs, Carlo Rubbia and the then young assistant professor Larry Sulak. The whole thing was chaotic, the experimentalists were arguing among themselves about whether they were seeing the sign of the new theory (neutral currents)—and again this problem of material reality and abstraction seemed riveting to me. By the end of 1978, I had a thesis topic—how experiments end, my own, slightly oddball entry into the philosophy of science.

3. How has your work offered original contributions to discussion on science? What does your work reveal that others fail to appreciate?

My central interest has been the collision of the most material things with the most abstract of ideas. It seemed to me, still does, that there are fundamental philosophical problems lurking in the company of material objects. Back in the early 1980s, I was fascinated by closure in the laboratory—how experimentalists actually made the decision to end an experiment in a world where the deductive endpoint of a rigorous mathematical proof was nowhere to be found. I wasn't interested in experimental confirmation of this or that famous theory—much less in the concocted philosopher stories about meter pointers, "degrees of confirmation," and the like. What intrigued me was a kind of shop floor epistemology—how physicists gain and secure knowledge in the messy, complicated world of getting things to work. I wanted to elicit the fabric of that laboratory epistemology—the problem of sorting out signal from the artifacts of apparatus or environment. I wanted to know how bits of theories, expectations and presuppositions shaped the way experimenters framed and addressed their lab work.—How researchers and collaborations built their views into devices themselves.

The part of my early work that I found most engaging was prying out the epistemic world of the experimenter from the better-known knowledge claim of finished theory. Along the way, I got fascinated with theory built into devices, abstractions made concrete, so to speak. I remember finding out that in one cluster of particle physics experiments I studied (those weak neutral current experiments I mentioned earlier that had emerged from a collaboration based at CERN and another with their locus at Fermilab), subgroups had battled it out. On the American side, one such subgroup kept turning a certain trigger on and another insisted on flicking it off. More particularly: one group installed an electronic configuration that would record events only if the recording device saw a muon (heavier version of the electron) in the particle collision. This was a legacy of the detector's tradition—such devices had famously been successful in a Nobel Prize winning demonstration by an earlier detector that certain kinds of neutrinos only produced muons, not electrons. The other group wanted if OFF: only then would it even be possible in principle to see the new "neutral" currents in which no muon issued from the particle crash.

At issue was this: in the 1970s, experimentalists wanted to push on the new unified field theories (the Weinberg-Salam-Glashow theory in particular) by addressing the prediction that there could be neutrino interactions that produced *no muons*. Now the historically successful trigger (recording only when there *was* a muon) would, by its very design, make absolutely invisible any of the events that might confirm this new unified theory. The experiment was built to ignore any events without muons—the neutral current events left no trace at all. That kind of phenomenon fascinated me. Traditions of experimentation on one side, traditions of theory on the other—with the two paced differently, moving with different rhythms, different assumptions, different standards of demonstration.

I wanted to press the inhomogeneity of physics practice even further. I wanted to know how the experimentalists related to the individuals and groups creating new kinds of instruments, I wanted to know about an epistemology of these chambers—cloud chambers that, with their builders, got altered into bubble chambers or into stacks of nuclear emulsions. Nuclear emulsion groups morphed in other ways into bubble chamber groups; electronic collaborations flattened Geiger counters into spark chamber plates. From these traditions of machines came an argument of sorts, but one utterly invisible to theoreticians: an epistemic tradition that prized high statistics *versus* those that valued the ability to produce events that could stand, at least putatively, on their own. One could ask epistemic questions of these machines in action beyond their specific applications to particular experimental questions. What were the least bits of information—sparks, chemical grains, bubbles, and droplets? How were these assembled into tracks, tracks into data, data into arguments, and arguments hammered into a form that could confront theory? I was interested in the structures of collaboration, of laboratory discipline, of industrial materials and scientific procedures—and importantly how all clattered against one other in the production of experimental knowledge.

This then was the picture I had by 1983 or so: physics as a tripartite assemblage of *intercalated* scientific subcultures—epistemic regimes that did not necessarily line up nicely to split into "paradigms," "programmes," or "epistemes." Put another way, philosophers of science had generally posed the question of continuity and coherence in physics as setting up in one of two ways. *Either* there was a single line of unbroken progression (for example the ever more inclusive reach of theory or the aggregation of empirical results as in positivism) *or* science divided into blocks (Newtonianism for example, or Einsteinianism). Thomas Kuhn—for example, he was by no means the only one—rejected the positivists' assumption that science proceeded by the aggregation of empirical results, and so concluded that science must divide into disjoint, incommensurable paradigms. My sense by the mid-1980s was that this was a wrongheadedly exclusive choice, and that what held the subcultures together was not that one of them was unbroken—all had their breaks—but instead, like Peirce's cable or Wittgenstein's thread, the overlaps gave the structure strength, not the existence of a golden chord.

All well and good. But just by insisting that the scientific subcultures had a good deal of autonomy one from the other, a new problem arose and got deeper as I pushed farther into the work toward *Image and Logic*. What held these subcultures together?

If experimentalists and theorists had different ontological commitments, different methods for assessment, different modes of publication, how did they actually speak to one another? How did language work at the boundary between languages?

For Quine, the problem had been that there were many modes of translation—no one could ever be singled out as privileged above all others. For Kuhn, the problem was rather the opposite: there wasn't even one truly adequate translation—between French and German or, as the case may be, between Einsteinian and Newtonian. But this all seemed fishy to me by 1989—a powerful generation of work in history of science had advanced by pressing hard on locality. In a sense, the single most significant task that my contemporaries and I were grappling with was *localization* of laboratory work, the need to avoid the mystical Zeitgeist arguments in which ideas were "in the air" floating from book to book like the medievals' notion of influence. And here we were still invoking Kuhn's *global* notion of language.

At last I had a question that I could formulate: what would be a *local* language that would correspond to *local* laboratory practice? This was in the fall of 1989-90—I was at the Institute for Advanced Study at Stanford—and I started grilling anthropological linguists about what they found—what empirically actually happened—at language borders. I didn't want to hear any more about languages of pure syntax, radical translation, or protocol languages. Instead, I began to read like mad about pidgins and creoles and in those coordinated, *local* languages, I began to find a way of formulating what happened at the boundary between physics and math, biology and chemistry or for that matter between experiment and theory. The key notion, it seemed to me, was that it was possible (even common) to coordinate very complex languages with *partial* and local coordinative structures. When a sea-going culture trades fish for a land-based culture's grain, the two do not have to agree on the global significance of the objects exchanged.

In particular, the two cultures do *not* need to square their wider and deeper cosomologies of fish, grain, and sustenance—all they need to do is agree on the appropriateness of the exchange. In this region of exchange, which I began calling the trading zone in the fall of 1989, one could begin to ask in some detail not just whether group X collaborated with group Y, but instead to ask precisely what the coordination was and how that coordinative structure developed over time.

All this took time to work out. By 1997, after the publication of *Image and Logic*, I had begun to turn to the third of the three subcultures, theory, but theory handled with the same approach as experimentation and instrument making, that is in terms of practices and partial coordination. I looked at Paul Dirac's hidden use of geometry, at Feynman's actual work at Los Alamos and how it shaped his views about rule-governed diagrammatic methods; at string theorists' creation of mirror symmetry and its hybrid position between field-theoretical physics and pure mathematics. My original intention had been to use these studies, with an analysis of clock coordination in special relativity, to form the third volume of the trilogy experiment/instrument/theory (*Theory Machines*). But clock coordination spiraled into a book on its own—*Einstein's Clocks, Poincare's Maps* (2003). Coordinated clocks provided a rich vein to mine as they stood so directly in the early twentieth-century triple intersection of the physics of electromagnetic theory and relativity; the map-making technologies of longitude determination; and the leading edge of a newly-forming philosophy of science.

This complex of ideas continues to fascinate me: scientific subcultures, their intercalated periodization, the trading zones in which material and abstract collide into one another, and the scientific interlanguages that bind them.

4. What is the relation between philosophy of science and scientific practice, science policy, or efforts for social justice? Can there be a more productive relation? Is this desirable?

It might seem as if philosophy of science is tied to politics and social justice by the thinnest of threads. But philosophy of science in its early, heady moment of Viennese logical positivism was, in fact, deeply and directly political. Its left wing (Rudolf Carnap, Otto Neurath, Philipp Frank) accepted, to varying degrees, the Austrian Marxism of Red Vienna. Indeed, the Vienna Circle more generally—along with allies in Berlin, the U.S., Britain, and elsewhere—saw itself as forming a picture of knowledge that would act as a counterweight to divisive and destructive nationalism and clericalism. Those years after World War I formed a heady, turbulent moment. Stripping philosophy of its obscurantism, putting the new sciences of perceptual psychology, relativity, physiology, and mathematical logic at the center of inquiry, made it possible to imagine a new status for an unphilosophical philosophy. It was supposed to be a hybrid movement that would open borders separating countries and movements as well as disciplines.

Carnap and Neurath launched their stripped-down protocol language ("Smell Ozone 12:00pm laboratory X") as a way of facilitating communication by removing meaningless utterances. It was to be a universal language—like Esperanto or occidental—but this time in the language of science: physical things, sensations, spacetime coincidences—and the logical connectives that bound them together. By the time Carnap, Neurath, and allies had formed their international movement for a unified science, it was explicitly a retort to the deluge of irrational, metaphysical, and hate-filled propaganda that floated around the rise of fascism. Enlisting luminaries like Bertrand Russell and Albert Einstein, the cause had a few momentary successes, but ultimately they were one more casualty of the Axis juggernaut. The Vienna Circle splintered, a few of its founders fled to England (including Neurath and Friedrich Waissmann), more to the United States (among whom included Carnap and Frank). A former student murdered Schlick on the steps of the university in an assault celebrated, politically, by fascists, as a triumph against Jews, secularism, and internationalism.

In the flight to the Anglophone countries, at one level the positivists achieved a success beyond their wildest dreams. Joined with indigenous movements in England and the U.S., logical positivism edged into logical empiricism, and by the 1950s, had become the accepted philosophy of both countries taught just about everywhere. At another level, a quadrant of the Vienna Circle fell off the boat on the voyage across Channel and Atlantic. Gone was Austromarxism, links to contemporary art and Bauhaus architecture. No more discussion of Freudian psychoanalysis—except, perhaps, to denounce it.

When politics re-entered the philosophy of science in the late 1960s and early 1970s, it was in a very different key. It was awkward to straightjacket the positivist/anti-positivist dispute into left-right terms, but such tailoring was in no short supply during those years. Paul Feyerabend and Thomas Kuhn took the Vienna Circle as their target. Carnap spent his life defending what he called the "principle of tolerance"—applied on the one side to questions of philosophy, and on the other in his life—at a time of much fury, he defended Angela Davis. Feyerabend's "Against Method" aimed at a too-rigid conception of *the* scientific method. Kuhn's injunctions against science-as-truth, his advocacy of disjunct, separately-validated paradigms (against a trans-historical protocol language), echoed with an egalitarian, anti-authoritarian orientation that many non-philosophers took up with gusto, some to defend a broad epistemic relativism within the sciences, others to question the truth claims of science relative to other modes of inquiry.

One of the most interesting arenas in which the anti-positivist ethos took root was within Social Studies of Knowledge during the 1980s—paradigms accepted, but now re-described as the tenets advocated by members of particular social groupings. Here Kuhn became a kind of patron saint (even if he didn't much reciprocate).

Using a too-narrow, purely intellectual historical lens, all this seems to make no sense—how did the left wing of the Vienna Circle—who paid for their politics very dearly, come to be the "right wing" target of the relativist philosophy of science of the late 60s? How did "positivist" come to be bandied about as if it stood for an absolutist branch of metaphysical realism, when Mach and his disciples always identified metaphysical realism as an arch-foe? The problem is that such an idea-affiliation account alone is not enough—not in the history of science and certainly not here. We don't have a proper history of HPS, not a substantial one in any case. But some of the elements of an account of the right-positivism/left-relativism split would have to reckon with at least two features. First, Vietnam-era, Cold War science was highly politicized. This was a time when, fairly or not, the prestige fields of the exact sciences—mathematics, computation, physics, chemistry—had come to be identified on campuses as conservative, if not implicitly pro-war. Second, the relativist stance toward scientific, para-scientific, and rejected-scientific accounts was seen as allied with the older, neo-Boasian relativism of anthropology, one that respected, valued, and sought to learn from the internal logic of cultures in their own terms, without ranking or dismissing one culture, language, or cosmology relative to another. The misfiring left/right battle inside the study of science came to be joined in one of the more bizarre flashpoints. After the fall of the Berlin wall, the Sokal affair pitted scientists, including those identifying themselves (like Sokal himself), with the New Left, and the left-identified science studies, against each other. No enemies to the left indeed.

Now squarely in a very different environment, the politics of science studies is changing. No doubt we'll have for a long time to come the symbolic politics of the older period. But something else is happening. Historians and other analysts of science—at least a few of them—have been in the courtroom and on the legislative floor. One example has been in the recent history of tobacco industry. Robert Proctor and Allan Brandt have testified (for the government and against the tobacco industry) in some of the largest legal battles ever waged. By showing just what the industry knew of tobacco danger—and when—their history has proven to be a key element in pushing back on one of the greatest killers in the whole of public health problems. Global warming is another battleground; and again historians of science entered the fray directly rather than purely symbolically. Naomi Oreskes (for example) conducted a much-discussed study for *Science* in 2004 showing that of the thousand or so refereed publications relevant to global warming, 75% argued that global warming was real and of human origin, 25% expressed no view, and *none* contested the warming/human origin stance. The so-called global-warming debate wasn't one. It was a fabrication, a piece of hype that satisfied warming-denying conservatives and infotainment all at once.

The importance of this shift from in-principle skepticism to an evaluation of where debate really lies, is crucial. When the cigarette industry famously quipped that "Doubt is our product"— and the climate change deniers seconded the slogan—a new politics of science enters the scene. It is still epistemological, but it now brings science studies to a certain form of specificity. Not "is science absolute or relative," but instead, "how does the epistemological rubber hit the road?"—"where is there doubt and in what proportion?," rather than "Is doubt possible?"

My own work at the border of politics and science lies closer to that of Brandt, Oreskes, and Proctor than to the more symbolic politics of classical HPS. One side of this work has been an attempt to join print and film. My first major project was on the moral-political fracture that split the physics community back in the late 1940s and early 1950s over whether scientists should build the hydrogen bomb. The goal of the film ("Ultimate Weapon: The H-bomb Dilemma," with P. Hogan, 2000), was to push hard on the conflict as it played out at the time—not to sanctify or martyr one side or the other, but instead to present a conjointly scientific and moral clash in terms that would trouble the idea of a reductive response. More recently, "Secrecy" (which I directed and produced with Robb Moss, 2008) aimed to get at the shifting alliance of security, democracy, and information. We are after the techniques of secrecy, the ways the securocrats invoke law to block access to information, the way policy and patterns of information flow have evolved hand-in glove. Conversely (in other work, jointly authored with Martha Minow), I've been after the way "technoprivacy" functions—the constant reconfiguration of privacy notions as the boundary conditions themselves shift with altering ideas of the self, and new forms of technical intrusion.

In some ways the history and practice of secrecy is the complement of the history of knowledge—and to understand how we grasp one is to understand a great deal about the other. At the broadest level, I think the history of science and science studies more generally, will increasingly grapple with the study of knowledge more generally—my guess is that the distinction between science and technology, or between science and social sciencewill increasingly strike us as beside the point. Why? Because the problems we face now—toxins, global warming, nano-scientific and genetic modifications of the genome (for example)—simply don't enter our world in the purity of laboratory models.

Part of the shift I've identified in these various forms is to move away from a purely symbolic politics and toward specific engagement; part too is a move to expand the address we put on the envelopes enclosing our work. Sometimes that may mean testifying in court as have Brandt and Proctor, sometimes answering questions posed by Congress, as has Naomi Oreskes. It may mean putting a film on television—or in film festivals, human rights meetings, National Laboratories, or theaters. How should I put this starkly? —The politics of philosophy of science has more interesting places to go than the re-warmed debate over whether positivism is right wing or left wing.

A second aspect of my work, this in print, has been to think of historiographic and philosophical work as tools, rather than Theory with a capital T. In this vein, I have thought about trading zones not as a one-size fits all account of how science works (there are areas of analysis where it is completely irrelevant), but as something that may help in understanding now, as well as in the past, what happens at the encounter point between groups addressing a common, or at least (commonly-disputed) domain. When algebraic geometers found that they had to talk to string theorists—a problem at the center of their discipline depended on it—the two groups did not homogenize. Mathematicians did not suddenly make rigorous all the lore that physicists had developed about Feynman diagrams, physicists did not suddenly start to reason in propositions, lemmas, and proofs. Far from it. Instead, they began to sort out a common exchange language in the border regions in which both groups had to work—they found certain ways of redescribing objects from the other side, they formed collaborations, held joint conferences, and began to plan ways of "training up" a new generation of physicist-mathematicians who could work more fluidly in the trading zone.

Some analysts, from a wide variety of domains, have begun to employ the ideas surrounding trading zones for a kind of epistemological politics. That is, they have used the notion to understand better what happens (for example) when soil scientists and farmers come to metaphorical blows over what is happening in the soil. Others have taken up the notion of these exchange languages and trading zones around the fisheries—as fishermen have to negotiate some form of common, *particular* understanding of what is happening in the nets and under the waves. I am hugely relieved to get away from the characterization of conflict regions like these as ones of "collaboration"—or worse yet a bygone punch-counterpunch between "relativists" and "absolutists."

2. and 5. What, in your view, are the most interesting, important, or pressing problems in contemporary philosophy of science as it relates to the broader study of science?

It would fly against everything I believe to militate for a single line of inquiry that "contemporary philosophy of science" should follow. I read with interest work in the foundations of quantum mechanical measurements, or for that matter on the contemporary sociology of the scientific workplace. History of philosophy seems to me to be at a good moment too, freer from ancient battles between those hunting the past for solutions to present concerns and those seeking to contextualize philosophical work. Of interest too is the confluence of interest in game theory that surrounds philosophy, economics, psychology—and more generally the philosophy of each of the special sciences. No, instead of preaching a line or lobbying for a theory, let me say a bit about where my own work is heading.

Here's what I think: As we expand focus from a narrow focus on scientific results to a broader scene that includes scientific practice, new questions emerge. Take the laboratory, where the history of experimental practices and skills have never stopped changing. In the 1920s, cosmic-ray physicists worked alone or in small groups of two or three; their required skill set included how to make the rather ornery Geiger-Mueller counters and cloud chambers. The statistics they employed were pretty simple—standard deviations, averages, not much more. Roll forward seventy-five years and the situation for their descendents, the particle physicists, was utterly different. Collaborations ran to 500 physicists—then 1000, and in the two contemporary collaborations at the CERN Large Hadron Collider, that number has spiraled north of 2500. Detectors zoomed past a million dollars back in the mid-1950s, and today can run a billion dollars, take a decade build, and occupy a volume the size of a factory building. Physicists specialize from early on—some do track-analysis software, others are experts on liquid calorimeters, yet others focus on cryogenics, vertex detectors, statistical analysis, visual display... or dozens of other specialties.

These contrasts are sociological, of course, but their epistemic consequences are deep and immediate. Arguments are assembled differently—and indeed the very process of coming to a conclusion has to proceed in new ways. There are prohibitions on who can speak for the collaboration (imagine 2500 different opinions brought to the press). Debates swirl over what someone has to do to become a signing author of a paper—at first, building seemed to be the sine qua non, then taking data, then, well, the kinds of validating contributions have multiplied alongside the differentiation of tasks in these sprawling virtual communities. Meetings, internal evaluations, formulae for determining how to referee—all this takes place in intranet discussions inside the collaboration itself.

Indeed, out of the large-scale structure of these scientific enterprises come new kinds of questions. Who, in this post-Foucaultian moment, counts as an author? Are these macro-collaborations forms of expanded cognition? What kind of evidence counts as convincing—and under what conditions? What are the circuits by which an argument must travel before it comes to be endorsed by the whole? I tried to get at some of these issues in *How Experiments End, Image and Logic,* and *Scientific Authorship.* But more generally, I've come to be increasingly interested in how the question "what counts as an *experiment*?" demands a simultaneous inquiry to the corresponding query, "what counts as an *experimenter.*" In other words, as we press the history of scientific practice, we put in play the historicity of what it means to be a scientist, and how those practiced values are imparted and cultivated.

A striking example of the confluence of science and scientist can be found in the history of scientific objectivity. In *Objectiv-*

ity (2007), Lorraine Daston and I showed that scientific objectivity has a history, and is not to be conflated cavalierly with other virtues we might want—accuracy, for example, or pedagogical utility, quantification, precision, replicability, or even truth. Now in ethics it is and has long been widely accepted that virtues may not pull in the same direction—it is hardly news to report that being fair and being just can come into conflict. By contrast, in epistemology we have been stunningly oblivious to such conflicting virtues: as if we expect (for example) that the accurate and the objective ought to lean in a similar direction. But they may not and, in fact, often don't. Just think of the many nineteenth century anatomists who preferred blurry, black and white, poor depth-of-field, pedagogically useless images because they were done with the chemical photograph,—the photograph could seem less vulnerable to personal artistry or taste.

Fixing attention on a remarkable genre—the scientific atlas—in writing *Objectivity* we found that across a wide swath of disciplines from physics and chemistry through pathology, anatomy and snowflakes, something happened in the period 1830-1870. Around that time, some scientists began, haltingly at first, to shift the aim of their representations of the basic working scientific objects. In the eighteenth century, the goal had been to construct the perfect object behind particular instances (ideal skull, ideal snowflake). For such a task, the best scientist was he who, in virtue of being a kind of sage, could intervene to part the curtains of mere experience. From the mid-nineteenth century forward, one began to see a very different ideal type of the scientific self. Scientists invoked as their highest ideal their *lack* of intervention, valuing above all self-abstention. Self-restrained registration of particulars became a virtue when it had been a vice; intervention, improvement, and idealization became vices when they had been virtues.

The emphasis on particular objects and self-restrained subjects altered the visual classification of working objects, marking a shift from (or at least a supplement to) the genial or sage-like scientific self. Self-restraint did double duty: it secured a kind of scientist and it vouchsafed the validity of a kind of science. In this sense, the history of the scientific self and the development of epistemology went hand in hand. Drawing a silhouette portrait line carves the foreground face from the background space—the demarcation defines both. Similarly, the line between subjectivity and objectivity at once secured *us* and stabilized the object. Early photographers were criticized for not being subjective enough; scientists at the same time could be blasted for being insufficiently objective.

My work now aims to push this reasoning further into different forms of scientific practice, to ask, as it were, how the scientific self both forms technical objects and is, in turn formed by them. An example: Ink blots have been around a long time—in the nineteenth century they formed a widespread parlor game: guests would look at a blot and speak to what they imagined the blot-forms to be. By the late 19th century, psychologists had seized the blots as a test—a test to measure the power of the specific faculty of the imagination (just as there were faculty-specific tests for memory, for calculation, for the ability to grasp at a glance, etc.). When Hermann Rorschach took up, systematized, and disseminated his blot-cards, he re-purposed them. From 1920 or so forward, the eponymous "Rorschach Test" lost its role as a training tool or probe of the power of the imagination. Instead, Rorschach configured his cards as if they were test patterns to be put before a candidate lens. The Rorschach Test offered a systematic way of characterizing the underlying, often unconscious modes of perception that that fixed, so to speak, the optics of perception: what do you suppress or emphasize in your perception of the world around you?

I want to know what had to be the case about the structure of the self such that the imagination-test could become a (Rorschach) perception test, how this latter test became even imaginable; and then, reciprocally, once the test became extraordinarily popular—applied millions of times for differential diagnosis, legal proceedings, and job placement—how did it, as a master metaphor, reshape the way see the scientific self?

The whole of this project, provisionally called *Building Crashing Thinking*, aims to follow a series of technological objects, and to pursue this back and forth. I want to ask first what the a priori of the scientific self has to be for particular technical objects to be imaginable at all; and then second to ask what the material a priori is for the reshaping of the scientific self. In a certain way, I suppose, my work always comes back to this theme that has fascinated me from the get-go. How do the altogether material objects of our scientific world collide with our most abstract concerns—concerns that may be mathematical-physical, concerns that may be political-moral, or concerns that may fall into the grounding strategies of epistemology?

References

d'Espagnat, B. *Conceptual Foundations of Quantum Mechanics*, 2nd edition. Reading, MA: W. A. Benjamin, 1971.

Holton, G. *Thematic Origin of Scientific Thought*. Cambridge: Harvard University Press, 1973.

Kant, I. *Critique of Pure Reason*. Translated by N. Kemp Smith. New York: St. Martin's Press, (1781) 1929.

Kuhn, T. *The Structure of Scientific Revolutions*. Chicago: University of Chicago Press, 1962.

Kuhn, T. *Black-Body Theory and The Quantum Discontinuity*. Chicago: University of Chicago Press, 1987.

Lakatos, I., and A. Musgrave (eds.). *Criticism and the Growth of Knowledge*. Cambridge: Cambridge University Press, 1976.

Marcuse, H. *One-Dimensional Man*. Boston: Beacon, 1964.

Oreskes, N. "Beyond the Ivory Tower: The Scientific Consensus on Climate Change." *Science*. 306(5702): 1686, 2004.

Quine, W. V. *From a Logical Point of View*. Harvard University Press, (1980) 1953.

Reichenbach, H. *Symbolic Logic*. New York: Macmillan, 1947.

Selected Bibliography and Filmography

Authored Books

How Experiments End. Chicago: Chicago University Press, 1987; Translation: French.

Image and Logic: A Material Culture of Microphysics. Chicago: University of Chicago Press, 1997; Translations (forthcoming): Korean and Spanish.

Einstein's Clocks, Poincaré's Maps. W.W. Norton: New York, 2003. Translations: Chinese, Czech, French, German, Greek, Hungarian, Italian, Korean, Portuguese, and Spanish.

Objectivity. (with L. Daston) Boston: Zone Press, 2007. Translation: German.

The Assassin of Relativity (in preparation).

Building Crashing Thinking (in preparation).

5. Peter Galison

Films

"Ultimate Weapon: The H-Bomb Dilemma." (producer with P. Hogan) premiered History Channel, 2000, 44m.
http://www.fas.harvard.edu/~hsdept/bios/galison-ultimate-weapon.html

"Secrecy." (director and producer with R. Moss), premiered Sundance Film Festival, January 2008. secrecyfilm.com

Edited Books

Big Science: The Growth of Large-Scale Research. Co-edited with B. Hevly. Stanford: Stanford University Press, 1992.

The Disunity of Science: Contexts, Boundaries, and Power. Co-edited with D. Stump. Stanford: Stanford University Press, 1996.

Picturing Science, Producing Art. Co-edited with C. Jones. New York: Routledge, 1998.

The Architecture of Science. Co-edited with E. Thompson. Cambridge: MIT Press, 1999.

Atmospheric Flight in the Twentieth Century. Co-edited with A. Roland. Dordrecht: Kluwer Academic Publishers, 2000.

Science in Culture. Co-edited with S. Graubard, and E. Mendelsohn. New Brunswick: Transaction Publishers, 2001; Reprint from *Daedalus,* Winter 1998.

Scientific Authorship: Credit and Intellectual Property in Science. Co-edited with M. Biagioli. New York: Routledge, 2003.

Einstein for the 21st Century. Co-edited with G. Holton, and S.S. Schweber. New Jersey: Princeton University Press, 2008.

Selected Articles

"Aufbau/Bauhaus." *Critical Inquiry.* 16: 709-752, 1990.

"Ontology of the Enemy." *Critical Inquiry.* 21: 228-266, 1994.

"The Collective Author." In M. Biagioli and P. Galison (eds.), *Scientific Authorship.* New York: Routledge, 2003.

"Images Scatter into Data, Data Gather into Images." In B. Latour and P. Weibel (eds.), *Iconoclash. Image Wars in Science, Religion and Art,* 300-323. Cambridge: MIT Press, 2003. Catalogue for the Exhibition Iconoclash, German Arts and Media Museum (ZKM), Karlsruhe, Germany, May 3 to August 4, 2002.

"Image of Self." In L. Daston (ed.), *Things That Talk: Object Lessons from Art and Science*, 257-296. New York: Zone Books, 2004.

"Mirror Symmetry: Persons, Values, and Objects." In M. Norton Wise (ed.), *Growing Explanations: Historical Perspectives on Recent Science*, 23-63. Durham and London: Duke University Press, 2004.

"Removing Knowledge." *Critical Inquiry*. 31: 229-243, 2004.

"Our Privacy, Ourselves in the Age of Technological Intrusions." (with M. Minow) In R. A. Wilson (ed.), *Human Rights in the 'War on Terror,'* 258-294. New York: Cambridge University Press, 2005.

"Ten Questions in the History and Philosophy of Science." *Isis*. 99: 111-124, 2008.

6
Ronald N. Giere

Center for Philosophy of Science
University of Minnesota, USA

---

1. How were you initially drawn to philosophical issues regarding science?

I was an elementary school student just after World War II when there was still a positive afterglow from the successful use of atomic bombs in the Pacific. I was drawn to the sciences and read many popular works about science, particularly physics, throughout high school. I ended up wanting to be nuclear physicist. I pursued this dream at Oberlin College. It being a liberal arts college, I was obliged to partake in other studies. I chose mathematics, of course, but also German Language and Literature and, most significantly, philosophy. I gave up the dream of being a physicist my first year of graduate school in physics at Cornell. The best students in the beginning graduate courses were Cornell seniors. I had already missed the boat. My second year I wrote an M.S. thesis with Philip Morrison on radiation from cosmologically remote sources, filling in the details of some calculations he had done on the back of an envelope. By then, philosophy being my second love, I was already applying to philosophy programs, ending up in the Cornell Philosophy Department, where there were practically no courses in the philosophy of science. Having just left physics, I made a conscious decision not to pursue topics in the philosophy of physics. I wrote my dissertation on prediction and confirmation with Max Black, but I really learned the philosophy of science during my early years teaching in the Department of History and Philosophy of Science at Indiana University. In sum, I was first drawn to the sciences. The philosophy of science came later. But the experience of having studied a science seriously before turning to philosophy has always influenced my approach to the philosophy of science.

2. What, in your view, are the most interesting, important, or pressing problems in contemporary philosophy of science?

Like many fields, the philosophy of science has become increasingly specialized during the last several decades. The first division is between studies focusing on particular sciences and studies of more general issues regarding the sciences as a whole. There is some overlap here as sometimes material from a special science is focused on more general issues, as one can study causation in quantum mechanics or in the social sciences. Reduction is major issue in the emerging philosophy of chemistry. Implications go both ways. Studies of special sciences are further specialized. Among philosophers of physics, some study quantum theory while others study space-time theories. Few do both. Even in the philosophy of biology there is now specialization between those focusing on evolutionary theory and those focusing on genetics. If one wants to know what are the major problems in these various areas, you have to ask specialists in those areas.

Studies of general issues are also further specialized. For twenty years following World War II, the range of topics was fairly limited: explanation, confirmation, causation, reduction, the nature of theories, and realism versus instrumentalism. That changed in the 1960s and 1970s as philosophers of science absorbed the impact of works such as Kuhn's *Structure of Scientific Revolutions* in the context of widespread cultural changes. The old issues did not go away, but new ones were added: scientific discovery, conceptual change, and the nature of scientific rationality. There now seems little interest in grand methodological theories of scientific rationality. These have been replaced by more particularized issues: experimentation, the roles of models in science, and scientific cognition. Of course each of these issues includes a range of subsidiary issues. Once again, specialists in each of these areas have their own ideas about the most important problems in their area.

The upshot is that I don't think it is now possible to pick out "the most interesting, important, or pressing problems in contemporary philosophy of science." The field is no longer sufficiently unified for that.

3. How has your work offered original contributions to discussion on science? What does your work reveal that others fail to appreciate?

For roughly the first twenty years of my career I continued the work begun in my dissertation. One of the major theses of the dissertation was that prediction and confirmation are not symmetrical. In some circumstances it matters to the strength of the evidence for a hypothesis that the confirming observations were actually predicted using that hypothesis rather than merely being true logical consequences of the hypothesis. The problem is then to characterize these circumstances.

The leading empiricist accounts of probability and induction at that time were those of Rudolf Carnap and Hans Reichenbach. And, indeed, these accounts implied that the temporal order in which evidence was obtained could make no difference in the strength of that evidence. Yet, I knew of many instances in the history of science in which it was widely thought by both scientists and commentators there is a difference. Moreover, many leading philosophical accounts of inductive reasoning from the 17th through the 19th centuries upheld the doctrine of there being a difference. Among supporters of the doctrine was C. S. Peirce, the founder of American Pragmatism, who advocated what he called "predesignation."

I concluded that there must be something fundamentally mistaken in the project of trying to understand inductive reasoning within the framework of modern logic, indeed, mostly in simple first-order logic. Following my inclination to look to the sciences, I decided to look at contemporary scientific work in probability and statistics. In this I was also inspired by Ian Hacking's early work. Both Carnap's and Reichenbach's theories of induction can be seen as rival accounts of statistical inference. I soon discovered that a version of Peirce's predesignation is a consequence of classical methods of statistical inference. As I continued my exploration of classical statistics I came to accept the common view of classical statisticians that what is called statistical inference is better understood as decision making, that is, deciding which hypothesis provisionally to take as approximately correct. This implies that there is no such thing as inductive inference that is an extension of deductive reasoning, or that even resembles deductive reasoning. In retrospect, I realized this was also Reichenbach's view, although in a much impoverished framework.

Classical statistics also agrees with Reichenbach that empirical probabilities are relative frequencies. Here, following Karl Popper, I was attracted to a propensity interpretation of physical probabilities. On this view, physical probabilities are a kind of weak causal connection that operates in individual processes. I thus became a modal realist, with propensities being a weak form of causal necessity. This was definitely against the tide of Humean sympathies at that time. In subsequent years, propensity interpretations have at least become respectable, although how widely accepted today is hard to gauge.

The move away from philosophers' toy models of induction to genuine statistical theory was more successful. By the mid-1980s, the few of us who had early championed statistical theory had been joined by a number of younger scholars such as Deborah Mayo. The subject that had earlier been known simply as "probability and induction" became "the foundations of probability and statistics." For me, this success was double edged. It meant that philosophers were taking seriously scientific work on probability and statistics. But it also meant that the subject was becoming increasingly specialized. Knowledge of statistical theory became prized for itself rather that for its connections with scientific reasoning in general. People in the field had difficulty relating to more general issues in the philosophy of science as well. They even became more and more detached from work in statistics itself. My personal reaction was similar to that I experienced regarding the philosophy of physics twenty years earlier. If I wanted pursue these issues in this way, I should have become a statistician.

I think my experience applies generally to the increasing specialization in the philosophies of the special sciences. Just as physicists don't understand current work in biology, philosophers of physics don't understand work in the philosophy of biology. Papers in these special fields are too often impenetrable to philosophical specialists in other sciences to say nothing about more general philosophers of science. Specialization in the sciences seems an inevitable consequence of scientific progress, but philosophers of science need not follow this path. In doing so, they run the risk of trying to meet scientists on the scientists' own turf. This is partly the result of a misunderstanding of how science works. A science is not just the sum of its theories, which can be learned from a textbook. A science is a distinct culture with its own norms and practices. They do not take kindly to strangers, that is, those without a Ph.D. in the science. As a result, specialists in the philosophy of a particular science end up talking mostly to each other, thus forming their own community with distinct norms and practices. It is, however, a virtue of philosophical studies of statistical inference that they at least have the potential for more general applications.

There was also a more internal reason for my abandoning a specialty I had helped create. During this period there was a growing movement, first among probability theorists and statisticians and somewhat later among philosophers, namely, Bayesian Inference. As it turns out, Bayesian inference is formally equivalent to Carnap's inductive logic. The difference is that, for Bayesians, probabilities are not logically determined, but, rather, regimented subjective degrees of belief. Thus, the many philosophers partial to Carnap's point of view found it easy to adopt a Bayesian account of statistical inference. Among statisticians, Bayesian methods are one useful methodological tool among others. Among many philosophers, such as the late Richard Jeffrey, Bayesian inference came to be seen as the foundation for all scientific inference and, in some cases, the foundation of all non-deductive reasoning. Some philosophical Bayesians began reconstructing historical cases within a Bayesian framework, with the predictable result that the hypothesis with the highest final probability turned out to be the historically preferred hypothesis.

From a naturalistic point of view, it always seemed to me completely implausible to think that scientists have been or now are intuitive Bayesians. This went against my own training in physics and my later observations of scientific work. Of course scientists use probability theory to calculate the probabilities of such things as the possible outcomes of a particular experiment, but never, I am quite sure, to calculate, even subconsciously, their own subjective degrees of belief in any hypothesis.

I had begun writing a much too ambitious monograph developing my own vision of classical statistics both for specialists and for a more general philosophical audience. In the end, I abandoned the project. I did not want to spend any more of my career engaged in defending a minority position in ever more technical terms. Besides, in a sense, I had already published my views, not in the form of a monograph, but in the form of a lower division textbook. Based on ten years of experience teaching a beginning course in scientific reasoning, the first edition of *Understanding Scientific Reasoning* was published in 1979. When the updated second edition appeared in 1984 I was already at work on a substantially revised third edition, which appeared in 1991. The early editions began with a brief introduction to the propositional calculus, and forms of scientific reasoning were structured as ersatz deductive arguments. In my teaching I had discovered that the logical framework did not help students understand the scientific reasoning behind conclusions and, in fact, got in the way of such understanding. I threw out the baby logic and went straight to the scientific cases.

The textbook, however, retained the spirit of an informal logic text in that it includes no discussion of rival understandings of statistical inference and, thus, no philosophical arguments. There is a lot of philosophy of science imbedded in the text, but it is not a philosophy of science textbook. I present my view as the way to understand and evaluate reports of scientific findings as they appear in the popular or semi-popular press. Thus, while reading a report regarding the effect of a new drug on some disease, one learns to ask crucial questions: Was the design of the study retrospective, prospective, or randomized prospective (an indicator of overall reliability)? Was the sample size large enough to detect small but important effects? Are reported differences statistically significant? But the text is also not an introduction to statistics. It employs a highly simplified version of classical statistics requiring no mathematical calculations beyond addition and subtraction, which one can do in one's head. There are rough rules of thumb that one can employ while reading a report. And there is even an easily internalized flow chart to guide one through an evaluation. In short, the goal of the text is to impart a kind of scientific literacy among non-scientists. I produced an updated fourth edition in 1997 but, frankly, could not face another update, so the changes in the 2006 fifth edition are due entirely to two co-authors recruited for the task. Besides, I had long since redirected my attention to other matters.

One should not abandon a major project without having other possible projects in mind. Typically for me, the inspiration for a new project came from outside philosophy. In 1973 I published a well-cited review of a volume of *Minnesota Studies* on relations between the philosophy of science and the history of science. It was entitled "History and Philosophy of Science: Intimate Relationship or Marriage of Convenience?" It is still cited in current discussions on this topic. In the review, I was much concerned about the tension between the normative ambitions of the philosophy of science and the descriptive practices of historians of science. I resolved the tension for myself a decade later, when, partly under the influence of the late Don Campbell, a philosophically minded social psychologist, I decided that the philosophy of science should be naturalized. The Quinian inspired title, "Philosophy of Science Naturalized," jumped out at me and I rushed to write a paper to go with the title. It appeared in 1985. My rush was obviously due my feeling that there were others, such as Richard Boyd, who were already moving in this direction and I wanted to be among the first to put that phrase prominently into print. At the time, naturalism in the philosophy of science was not highly regarded. Twenty years later it is a widely held, perhaps even the dominant position, in the philosophy of science.

If one is to naturalize the philosophy of science, the question immediately arises, "To what are you going to naturalize it?" One option was the social world, as advocated by a then newly radicalized sociology of science. Another was the growing field of cognitive science. The message coming from "new wave" sociology of science was that what theories scientists settle on may be wholly dependent on social interactions. How the world might be may play no role. That was far too radical a view for me. But I was impressed that sociologists, like historians, were talking about real scientific practices rather than only established theories, which were then the standard focus of inquiry in the philosophy of science. I had the feeling that people who then took the new sociology of science at all seriously, such as Ian Hacking and Tom Nickles, were regarded with suspicion in the field at large, even if they were critical. I chose cognitive science as reflected in the title of my 1988 book, *Explaining Science: A Cognitive Approach*. But I did not exclude social aspects of science.

From my point of view, the overall naturalistic project of the book, announced in the first chapter, was at least to outline some elements of an answer to this question: How can we explain how humans, with just their evolved cognitive capacities and social structures, have managed to construct modern science? How could humans come to know such things as the structure of DNA or the age of the universe? The answer, of course, must have many components, cognitive, social, historical, and who knows what all else. But the focus must be on distinct mechanisms, not by appeal to high level methodological principles.

The cognitive orientation of the book immediately put me in the company of a few others, such as Nancy Nersessian and Paul Thagard, who were then developing what became "The Cognitive Study of Science." My subsequent participation in this enterprise has been sporadic, but the book is often cited in this context and I continue to receive invitations to contribute to volumes or partake in conferences on the subject. Cognitive studies of science are not now as well appreciated within the philosophy of science community as I had hoped. They are, somewhat ironically, better regarded in the wider science studies community. I wrote the entry on cognitive studies of science and technology for the 2008 edition of *The Handbook of Science and Technology Studies*.

By far the most cited part of *Explaining Science*, down to the present day, is the third chapter on models and theories. Significantly, as several critics pointed out, there is nothing particularly cognitive in this chapter. It is all about the important roles of models in science. I learned about models from Bas van Fraassen in 1968 when he was my office neighbor while visiting at Indiana, and I have retained a model-based view of scientific knowledge ever since. I continue to differ with him on the issue of realism. Chapter four of *Explaining Science* is titled "Constructive Realism" in direct contrast to his "Constructive Empiricism." And my view of the nature of models has since drifted considerably from his logic-based approach.

In those days, a model-based view of theories (then usually called "the semantic view of theories") was definitely a minority view. It was championed in the previous generation by Pat Suppes. In my generation, other early advocates were Joe Sneed (Suppes' student), Fred Suppe, and, of course, van Fraassen. Over the years we have been joined by many others, including especially Nancy Cartwright and Paul Teller. Forty years later, a model-based view may now be the majority view. Indeed, the semantic view, including the view in *Explaining Science*, is now often presented as the new "received view" by people such as Margaret Morrison who seek to go beyond it in various ways.

During the next decade I was much involved directing the Minnesota Center for Philosophy of Science and helping to run the Philosophy of Science Association, including two year stints each as future, present, and past President. In the latter roles I presided over a major reorganization of the Association. I had no time for another book, but did continue writing papers, some of which were collected in a 1999 volume provocatively titled *Science without Laws*. My original title went on: *Realism without Truth, and Judgment without Rationality*. That title did not get past the marketing department at the University of Chicago Press. And, of course, it was somewhat ironic. What we can do without is the philosophical analysis of laws as true universal generalizations, or, worse yet, relations among universals. Similarly, what science seeks is not simple truth, but models that fit well enough for current purposes. And, finally, we should give up looking for *a priori* principles of rationality and settle for decision procedures with desirable properties relative to current goals.

In the conclusion of *Science without Laws* I announced a new project, arguing that scientific knowledge, both experimental and theoretical, is perspectival in the way that human color vision is perspectival. I suggested a perspectival realism that is a further elaboration of constructive realism. An elaboration of this view was published in 2006 under the title, *ScientificPerspectivism*. I have recently seen several books quite independently advocating a version of what I call scientificperspectivism, and at least one international conference has been held on the general topic. Especially significant for me is van Fraassen's exciting 2008 book, *Scientific Representation: Paradoxes of Perspective*. Of course he advocates a "perspectival empiricism." It will take at least a decade to learn how influential these various notions ofperspectivism might be.

In the final chapter of *ScientificPerspectivism*, I introduce the idea of distributed cognition. One of the ways science has advanced is through the construction of various artifacts, both conceptual (the calculus) and material (the telescope). It is enlightening to think of humans together with such artifacts as comprising a larger cognitive system that products scientific knowledge. Scientific cognition is for the most part now distributed cognition. The connection withperspectivism is that introducing new cognitive artifacts is one way of changing our perspectives on the world.

4. What is the relation between philosophy of science and scientific practice, science policy, or efforts for social justice? Can there be a more productive relation? Is this desirable?

I think that a major task for general philosophy of science should be to develop a systematic meta-view of all important aspects of scientific practice. Such a view could be helpful to scientists in gaining a better overall understanding of their own practices, if one takes pains to present it to working scientists in a way they might appreciate. Working scientists naturally have their own meta-views about what they are doing, but these are rarely systematic since scientists have little time for such things. Often scientists' meta-views simply reflect an opportunistic adoption of some earlier high profile philosopher of science, such as Popperor Kuhn.

Something similar holds for uses of the philosophy of science in considerations of science policy or other social projects. The better one understands how sciences work, the better will be applications in other areas. There is hardly any way to consider broader applications if one is a philosophical specialist in some particular science and perceives their job to operate in parallel with scientists at the level of scientific theory. This does not necessarily apply to philosophers operating at a less specialized level. A good example here is philosophers of biology, such as Philip Kitcher and Michael Ruse, who have made great efforts to bring a sound understanding of evolutionary theory and human genetics to an audience well beyond other philosophers of science. Other good examples include Nancy Cartwright's work on science policy issues in the social and bio-medical sciences and Kristin Shrader-Frichette's more general work on science policy. Finally, feminist philosophers of science, such as Helen Longino, continue to apply their understanding of the philosophy of science to all sorts of issues of special concern to feminists.

5. Where do you see the field of philosophy of science to be headed? What are the prospects for progress regarding the issues you take to be most important?

Before the advent of computer models of weather patterns, it was often said that the most reliable prediction about tomorrow's weather is that it will be like today's weather, whatever that might be. That maxim would seem to apply to almost any academic field. It is hard to spot new developments before they happen. Nevertheless, I will hazard a few guesses.

One area that seems to me sure to grow concerns the influence of computers on scientific practice. In the past, philosophers of science, with a few exceptions such as Fred Suppe and Paul Humphries, have in general been slow to recognize just how much computers have changed the way science is done. In some cases, the advent of high-speed computing has opened up whole new areas of science which could not otherwise be pursued. Examples include medical imaging, climate models, and a host of other simulation models which are now being studied by a number of younger scholars. I think it not too much of an exaggeration to say that computers are the most influential methodological tool developed since statistics, and maybe even since the invention of the calculus and analytic geometry.

Within the philosophy of biology, I think we will see increased interest in genetics, as exemplified, for example, in the recent work of Ken Waters. The modern philosophy of biology has its origins in the 1960s when, thanks to N. R. Hanson, Kuhn, and others, the history of science was having a great impact on the philosophy of science. This is reflected in the fact that many of the leading philosophers of biology today have strong interests in the history of biology. Yet, although the most recent revolution in genetics began with the discovery of the structure of DNA in the 1950s, except for questions about the reduction of biology to chemistry, philosophers of biology since the 1960s, with some notable exceptions, have focused mainly on evolutionary theory. Given the success of the human genome project, my guess is that philosophical interest in genetics will now get its due.

Turning to things closer to my own concerns, I think that the current interest in the role of models in science, and in distinctions between models and theories, will continue for some time. The full impact of thinking of science in terms of models has yet to work its way through the philosophy of science. I am particularly interested in agent-based accounts of representation, a three place relation in which agents use both physical and symbolic resources to represent aspects of the world. There are also connections between this topic and the cognitive study of science, as in Nancy Nersessian's work on model-based reasoning.

There are also connections between naturalism and Pragmatism that I would like further to explore. The main connection for me has been that naturalism is necessarily non-foundational. It is also a primary doctrine of Pragmatism that there are no ultimate foundations. One can only begin with the beliefs one has and proceed by subjecting actual conflicts among one's beliefs to critical inquiry. Anything can be questioned and subject to inquiry, but not everything at once, contrary to the Cartesian strategy.

Finally, I would like to mention an old interest that I have only recently taken public: the scientific study of religion. I have long been interested in relations between science and religion, and for several years taught a freshman seminar on the topic. My main concern, however, has been the question of why religion continues to play such a big role in an advanced technological society such as the United States when, by modern scientific standards, the ontological claims of religions are without any respectable empirical foundation. I have also become increasingly alarmed by the ever growing role of religion in U. S. public life for the past quarter century, extending even to the highest levels of government.

I was overjoyed when Dan Dennett recently published *Breaking the Spell: Religion as a Natural Phenomenon*. Not only did he introduce me to a vast literature I was only beginning to uncover, he helped make the subject respectable among philosophers. It took almost no effort to transpose his interests into a philosophy of science context. There are at least half a dozen scientific disciplines that include people focusing on finding scientific explanations of various aspects of religion, but there is little interaction among them. This presents special problems for philosophers of science who mainly study single mature theories. There is ample room here for the development of a new area of study within the philosophy of science, but it faces the considerable difficulty that, right now at least, it can only be pursued by people at full rank or recently retired. No one is currently likely to be tenured or promoted by a philosophy department for publications in this area. Likeperspectivism, it will take at least a decade to see if this enterprise succeeds. I hope I am still around to find out.

References

Dennett, D. *Breaking the Spell: Religion as Natural Phenomenon*. Penguin, 2006.

Kuhn, T. *The Structure of Scientific Revolutions*. Chicago: University of Chicago Press, 1962.

van Fraassen, B. C. *Scientific Representation: Paradoxes of Perspective*. Oxford University Press, 2008.

Selected Bibliography

Books

Understanding Scientific Reasoning. New York: Holt, Rinehart & Winston, 1979; Second Edition, 1984; Third Edition, 1991; Fourth Edition, 1997; Fifth Edition, 2006.

Explaining Science: A Cognitive Approach. Chicago: University of Chicago Press, 1988.

Science without Laws. Chicago: University of Chicago Press, 1999.

ScientificPerspectivism. Chicago: University of Chicago Press, 2006.

Selected Articles

"History and Philosophy of Science: Intimate Relationship or Marriage of Convenience?" *The British Journal for the Philosophy of Science.* 24: 282-297, 1973.

"Philosophy of Science Naturalized." *Philosophy of Science.* 52: 331-356, 1985.

"Scientific Realism: Old and New Problems." *Erkenntnis.* 63(2): 149-165, 2005.

"Perspectival Pluralism." In *Scientific Pluralism*, S. Kellert, H. Longino, and C. K. Waters (eds.), *Minnesota Studies in the Philosophy of Science, Vol. XIX.* University of Minnesota Press, 2006.

"Modest Evolutionary Naturalism." *Biological Theory.* 1(1): 52-60, 2006.

"Cognitive Studies of Science." In E. J. Hackett, O. Amsterdamska, M. Lynch, and J. Wajcman (eds.), *The Handbook of Science and Technology Studies*, 259-278. MIT Press, 2007.

"Models, Metaphysics and Methodology." In *Nancy Cartwright's Philosophy of Science*, S. Hartmann, C. Hoefer, and L. Bovens (eds.). Oxford: Routledge, 2008.

"Why Scientific Models Should not be Regarded as Works of Fiction." In *Fictions in Science: Philosophical Essays on Modeling and Idealization*, Mauricio Suárez (ed.), 248-258. Routledge, 2009.

7
Adolf Grünbaum

Andrew Mellon Professor of Philosophy of Science
University of Pittsburgh, USA

---

1. How were you initially drawn to philosophical issues regarding science?

In Cologne, Germany, where I was born in 1923 (thus the German Umlaut on my last name), when I was about 12 or 13 years old I ran across an eclectic book in German entitled *"Philosophie"* (*"Philosophy"*), which was part of a German series "Kultur der Gegenwart" (*"Present-day Culture"*), also featuring similar books in other fields. As I read in the Philosophy book about speculative ideas on cosmogony and cosmology, for instance, it became clear to me that the results of the natural sciences were fundamentally *indispensable* to arrive at evidentially well supported answers to the major questions in that area. Relatedly, I resolved to study physics and mathematics as prerequisites for the systematic study of philosophy, with a major focus on the philosophy of science.

Accordingly, as an undergraduate at Wesleyan University (in Middletown, CT.), I took a double major in mathematics and philosophy, and in Graduate School at Yale University, I took an M.S. degree in physics en route to a Ph.D. in philosophy, with a doctoral dissertation on "Modern Science and Zeno's Paradoxes" under the renowned philosopher of science Carl G. Hempel.

After I had begun graduate study in philosophy at Yale, I complained that the departmental faculty did not have *any* representative of the logical empiricist point of view, which had originated in the Vienna Circle but was then in its heyday in the USA. In fact, the Yale Philosophy Department had refused to give tenure to its then Assistant Professor Charles Stevenson, author of *Ethics and Language*, who was a representative of the emotivist conception of ethics that had been associated with logical empiricism. Assistant Professor of Philosophy Monroe Beardsley took it upon himself to tell the then Chairman of the Department, Professor Brand Blanshard, that, unless it appoints a logical empiricist, I would leave for another graduate department elsewhere *and* would take a group of fellow graduate students with me. But Beardsley's warning was conjectural, since I had not issued it.

Yet it worked: The Department promptly invited the eminent Carl ("Peter") Hempel to present a paper. At the time, he was teaching at Queens College in New York City. He was both a recognized exponent and a developing critic of his logical empiricist patrimony. Immediately after Hempel had delivered his talk in the departmental Seminar, Prof. Blanshard came over to me in the back of the room to get my reaction, which was glowing. Thereupon, Yale offered Hempel a tenured Associate Professorship, and—to my delight—he then joined its Philosophy Department. Hempel had been a Ph.D. student of the very eminent Prof. Hans Reichenbach at the University of Berlin, whence both of them left Germany to escape the Nazis. Reichanbach's epochal 1928 treatise *Philosophy of Space and Time*, in its German original, and his other prolific writings, were the strongest influence on me in my own work in space-time philosophy *and* in the philosophy of physics more generally.

I met Reichenbach and his wife Maria repeatedly, but I never studied with him, unlike Wesley Salmon and Hilary Putnam.
Though Hempel was Reichenbach's doctoral student, Hempel did not work in the technical philosophy of physics as such. Thus, Reichenbach was my intellectual grandfather, as it were. And I have enjoyed a life-long warm friendship with his widow Maria, who worked indefatigably to bring his writings to the public in both English and German. But, of course, Hempel left a very beneficent and strong intellectual imprint on me both in his superb graduate courses and as a valuable sounding board for the ideas I developed autonomously in my Ph.D. Dissertation. He was a lovely, benevolent, very kind and guileless man, and it was a sheer joy to work with him! After I had been his Ph.D. student, both he and his wife Diane became very affectionate friends of both my wife Thelma and myself until they each died. They both lived into their nineties.

Although the Yale Philosophy Department had been eager to recruit Hempel, it dragged its feet in promoting him to a full professorship, apparently because his publications at the time were not "between hard covers." This soured him on the Department, and although it finally granted him the promotion, he went to

Princeton when offered a full professorship there. Happily, after he retired from Princeton, he joined the faculty of my University of Pittsburgh as a University Professor of Philosophy. He served for eight years, and retired completely only when diminishing eye sight compelled him to do so. His wife Diane emphasized that these eight years had been the happiest of their life in the academic world.

2. What, in your view, are the most interesting, important, or pressing problems in contemporary philosophy of science?

In present-day culture, in which the theory of biological evolution and the achievements of the Enlightenment are under attack by fideist religious vigilantes, philosophers of science in colleges and universities cannot afford to focus only on issues internal to the philosophy of science. Their work should contribute to the public understanding of the role of critical thinking in the validation of major theories in the natural and social sciences. Significantly, Oxford University in England has a Chair in the Public Understanding of Science, currently held by the evolutionist Richard Dawkins, author of the 2006 best seller *The God Delusion*. This much is my answer to your question as to "the most pressing problems" in present-day philosophy of science. As for "the most interesting and important" problems in present-day philosophy of science, I am reluctant to say, since my valuations are perhaps too laden with subjectivity.

I do believe that philosophers of science should engage in public debates with the proponents of intelligent design, especially when they try to jeopardize the teaching of evolution, both biological and cosmogonic, in the public schools. Challenging them does not have the downside of inflating their importance, since they are already very vocal and need to be undermined. I do not see good evidence that Dawkins's combativeness is counter-productive. And I believe, as he does, in fighting fire with fire.

3. How has your work offered original contributions to discussion on science? What does your work reveal that others fail to appreciate?

I would feel much more comfortable, if you put that sort of question about my work to others who are familiar with it. But I would be falsely modest, if I did not report that, for example, far and wide, both critics and proponents of (Freudian) psychoanalytic theory have hailed my critique of that theory, and of its claim to scientificity, as by far the most important *ever* in over a century since Freud first founded psychoanalysis. Here I can refer you to my lengthy essay "The Reception of my Freud-Critique in the Psychoanalytic Literature," which appeared in the July 2007 issue (vol. 24, pp.545-576) of the journal *Psychoanalytic Psychology*, the official journal of the Division of Psychoanalysis of the American Psychological Association. I have distilled the gravamen of my challenge to Freudian psychoanalytic theory and therapy in my article "Critique of Psychoanalysis," which appeared in E.Erwin (ed.), *The Freud Encyclopedia: Theory, Therapy & Culture*, New York, Routledge, pp.117-136. Perhaps I should mention that, in 1990, Yale University gave me the Wilbur Cross Medal "for outstanding achievement."

4. What is the relation between philosophy of science and scientific practice, science policy, or efforts for social justice? Can there be a more productive relation? Is this desirable?

We know from the testimony of towering figures like Einstein that philosophy of science considerations played a significant role in their scientific creativity and practice. The rest of your 4th question is far too huge for an answer in an interview, and some of it exceeds my competence.

5. Where do you see the field of philosophy of science to be headed? What are the prospects for progress regarding the issues you take to be most important?

Fundamentally, I believe that there is too much unpredictability for reliable answers to what you ask. I cannot speculate regarding the queries under this rubric.

The interested reader may wish to supplement *all* of my answers here by reading my lengthy chapter "Autobiographical-Philosophical Narrative" in A. Jokic (ed.), *Philosophy of Religion, Physics, and Psychology: Essays in Honor of Adolf Grünbaum*, Prometheus Books, Amherst, NY, 2008.

References

Dawkins, R. *The God Delusion*. Boston: Houghton Mifflin, 2006.

Stevenson, C. *Ethics and Language*. New Haven: Yale University Press, 1944.

Reichenbach, H. *The Philosophy of Space and Time.* New York: Dover, 1958.

Selected Bibliography

Selected Books

Philosophical Problems of Space and Time. New York: Alfred A. Knopf, Inc., 1963.

Modern Science and Zeno's Paradoxes. Middletown, CT: Wesleyan University Press, 1967; Revised edition, London: George Allen and Unwin Ltd., 1968.

Geometry and Chronometry in Philosophical Perspective. Minneapolis, MN: University of Minnesota Press, 1968.

Revised edition, Boston Studies in the Philosophy of Science, vol. 12. Dordrecht, Netherlands: D. Reidel Publishing., 1973.

The Foundations of Psychoanalysis: A Philosophical Critique. Berkeley, CA: University of California Press, 1984.

Validation in the Clinical Theory of Psychoanalysis, A Study in the Philosophy of Psychoanalysis. Madison, CT: International Universities Press, 1993.

Selected Articles

"The Duhemian Argument." *Philosophy of Science.* 27(1): 75–87, 1960.

"Free Will and Laws of Human Behavior." *American Philosophical Quarterly.* 8(4): 299–317, 1971.

"Is Freudian Psychoanalytic Theory Pseudo-Scientific by Karl Popper's Criterion of Demarcation?" *American Philosophical Quarterly.* 16(2): 131–141, 1979.

"The Placebo Concept in Medicine and Psychiatry," *Psychological Medicine.* 16: 19-38, 1986.

"The Pseudo-Problem of Creation in Physical Cosmology." *Philosophy of Science.* 56(3): 373–394, 1989.

"Pseudo-creation of the Big Bang." *Nature.* 344(6269): 821–822, 1990.

"Creation as a Pseudo-Explanation in Current Physical Cosmology." *Erkenntnis.* 35: 233–254, 1991.

"A Century of Psychoanalysis: Critical Retrospect and Prospect." In M. S. Roth (ed.), *Freud: Conflict and Culture*, 183-195. New York: Alfred A. Knopf, 1998.

"Theological Misinterpretations of Current Physical Cosmology." *Philo.* 1(1): 15–34, 1998.

"Critique of Psychoanalysis," in E. Erwin (ed.), The Freud Encyclopedia. New York; Routledge, 117-136, 2002.

"The Poverty of Theistic Cosmology." *The British Journal for the Philosophy of Science.* 55(4): 561-614, 2004.

"Why is there a Universe AT ALL, Rather Than Just Nothing?" In C. Meister, and P. Copan (eds.), *The Routledge Companion to the Philosophy of Religion*, 441-451. London and New York: Routledge, 2007.

Feschriften

Physics, Philosophy and Psychoanalysis: Essays in Honor of Adolf Grünbaum. R.S. Cohen, and L. Laudan (eds.). Lancaster: Reidel, 1983.

Philosophical Problems of the Internal and External Worlds: Essays on the Philosophy of Adolf Grünbaum. J. Earman, A. Janis, G. Massey, and N. Rescher (eds.). Pittsburgh: University of Pittsburgh Press,1993.

Philosophy of Religion, Physics, and Psychology: Essays in Honor of Adolf Grünbaum. A. Jokic (ed.). Buffalo: Prometheus Books, 2009.

8
Sandra Harding

Graduate School of Education and Information Studies
University of California, Los Angeles, USA

---

1. How were you initially drawn to philosophical issues regarding science?

I was initially drawn to philosophy of science through several kinds of experiences. For one thing, I so enjoyed my graduate classes at New York University in the late 1960s with Richard M. Martin, who had been a student of Rudolf Carnap's. He conducted his classes entirely through conversations with us about what we reported that we didn't understand or found problematic in the readings for the day. Other students seemed to want him to provide well-outlined lectures. But I found his style of teaching most stimulating. It was so interesting to hear what other students found incomprehensible or problematic. And such conversations with me each week created an on-going one-on-one dialogue that stimulated my even deeper engagement with the class issues and readings. Moreover, I was fascinated by Martin's personal relationship to logical positivism. In terms of classroom ritual as an introduction to a discipline, it's hard to beat having a moment of silence in class each year to commemorate Carnap's birth date.

Which leads me to a second pull toward philosophy of science: the intellectual excitement at that time over the beginnings of the break from logical positivism. A student in Martin's class in 1968 or 69 asked about the significance of Kuhn's *The Structure of Scientific Revolutions*, which I had already had the opportunity to discuss in an advanced political philosophy seminar that I was auditing at SUNY Albany the year before I entered graduate school in philosophy. Martin told us not to bother reading it: it was a "piece of ephemera" that would soon pass away. Of course the rest of the class immediately went out and bought the book and we then all discussed it endlessly in the NYU coffee shops.

I also took a couple of courses on Quine's *Word and Object* including one that also focused on Polish logic's metaphysical assumptions and how these differed from the standard ones we had been taught in our courses on formal logic. It was so strange and exciting to imagine that there could be alternative assumptions to those characteristic of what we had been taught was the foundation of any and all rational thought. What were the consequences of this odd phenomenon? The world seemed to imperceptibly shift. This last course was with Henry Hiz, the linguist/philosopher who was visiting NYU from the University of Pennsylvania, and who had been Noam Chomsky's teacher. Meanwhile, we had been reading Chomsky's attack on empiricism, which he related to his criticisms of the U.S.'s involvement in the Vietnam War, though this was not a topic discussed by our faculty, at least not in my experience.

Then I wrote my dissertation on Quine's epistemology and its only partial break from logical positivism. This was before anyone thought Quine had an epistemology, for he was at that point treated primarily as a logician and metaphysician. So the opportunity to think in the context of widespread controversies about conventional philosophies—controversies which seemed to have social relevance—certainly pulled me toward philosophy of science.

With this bright and shiny philosophical training in hand, I went to my first teaching job in 1973, which was at an interdisciplinary experimental social sciencefocused college (a part of SUNY Albany) in which most of the other 17 faculty had been happy participants in the 1960s upheavals at Berkeley and elsewhere. My new physicist and political science colleagues begged me not to "pollute the young minds" in our classes with the thinking of Carnap, Reichenbach, or even Quine. They passed me a purple mimeographed copy of Feyerabend's *Against Method*; one of the physicists had taken classes with him at Berkeley. They attended the philosophy of science classes I taught and vigorously contributed Marxian questions about science, its dominant philosophy, and their social functions. And they passed me a mimeographed copy of a paper by the Canadian feminist sociologist of knowledge, Dorothy Smith. This was her first in which she began to ask questions about how to produce sociological research that was "for women," not just about or by them ("the standpoint of women")—a topic on which I subsequently have devoted considerable energies (Smith, 1987; reprinted in Harding, 2004).

Thus during my graduate education and early teaching, I became used to thinking about alternative intellectual systems (Polish vs. standard logics, Quine, Kuhn, and Feyerabend, and Marxist approaches to science vs. the more strictly analytic philosophy which flourished in the 1960s and 70s). I also became used to thinking about what different disciplines and political orientations cared about with respect to philosophies of science, and about the political relevance of sciences and their philosophies. Controversy seemed to me to sharpen philosophic insights and analyses. And I became accustomed from the beginning of my teaching to teaching philosophic issues to students from many disciplines, and to having scholars from other disciplines in my classrooms. These experiences would shape the directions my work took in subsequent decades as well as the institutional locations which I have found most fruitful for my own thinking.

Meanwhile, the women's movement was beginning. The Society for Women in Philosophy's first meetings were held in 1972 and 73. My feminist colleagues who were trained in ethics and political philosophy were producing ground-breaking challenges to conventional work in those fields. What could someone like myself, working in epistemology and the philosophy of the natural and social sciences, contribute to the gathering intellectual and political storm that the women's movement was creating by the mid 1970s?

By 1977 or so, the late Merrill Hintikka and I were talking about a project which addressed this issue. It became *Discovering Reality: Feminist Perspectives on Epistemology, Methodology, Metaphysics, and Philosophy of Science*. This was originally planned as a special issue of *Synthese*. We put out a call for papers, expecting that we would get a dozen or so submissions and be able to select three or four, to which we would add two already in print but relatively unknown at the time, and have a very nice little journal issue. However, when we received well over one hundred submissions, Jaakko Hintikka gave us the gift of inviting us to publish a larger selection of them as a volume in the Synthese Library series.

Not all of the philosophers on the Synthese Library editorial board were happy to hear about the topic of this proposed volume. One purportedly e-mailed his colleagues that he supposed the cover would feature a picture of Betty Friedan triumphantly waving her bra. Perhaps it is not too farfetched to suggest that the book itself performed such an act in an intellectual mode! That collection was arguably the first such volume devoted to formulating specifically feminist philosophy of science questions.[1] That is, it began to envision what epistemology, metaphysics, methodology and philosophy of science could look like if they started off with questions arising from women's lives. Publication by Kluwer turned out to be even more of a gift than we initially could have imagined. The book was read all over Europe and the other parts of the world where Kluwer books sell. Moreover, it kept selling long beyond when one might expect an anthology to prove marketable. It was astonishing to me that Kluwer requested a 20th anniversary, second edition of this collection, which appeared in 2003. How many other two-decade old anthologies have found a second edition publication? This phenomenon speaks to the power of the feminist epistemology and philosophy of science which developed in the intervening twenty years.

By the late 1980s, work by women of color was on the frontiers of feminist thinking. I could no longer justify teaching a course merely on gender and science, and began to try to locate writings that addressed the distinctive effects Western sciences had had in the past and continued today to have on women of color—from the development in the Nineteenth Century of scientific racism, to the 20th Century eugenics movement, and the 1970s pharmaceutical companies' testing of contraceptives on Puerto Rican women before they were regarded as safe for U.S. white women. I was also beginning to participate in the emergingpostcolonial Science and Technology studies movement—of which more below. Editing *The 'Racial' Economy of Science* (1993) emerged from my exploration of the often hard-to-locate literatures on these topics.

I was forced to begin to think more deeply about what was wrong with contemporary Western sciences and their philosophies. Clearly they suffered from the kind of epistemological undevelopment that these analyses revealed. These sciences and their philosophies had little or no resources that could enable them to recognize that their very best work, not just individually aberrant examples of "bad science" or "bad philosophy," was deeply embedded in frequently politically regressive social politics. It advanced social progress only for the few, already overadvantaged, citizens

---

[1] By 1970 feminists in biology and medical sciences had begun raising philosophic questions about the objectivity of these scientific fields in the course of their critical analyses—e.g., The Boston Women's Health Collective, *Our Bodies, Ourselves*, had appeared in 1970; the "Genes and Gender" series edited by Ethel Tobach and Betty Rosoff had began appearing in 1978.

of the world contrary to conventional rhetorics about how scientific knowledge was itself a social good. Moreover, this containment by politically regressive politics undermined its empirical and theoretical adequacy, not to mention its claims to value-neutrality.

2. What, in your view, are the most interesting, important, or pressing problems in contemporary philosophy of science?

Four problems interest me these days. Each is generated by reflection on the current contexts in which Western sciences and their philosophies function.

First, in general, philosophies of science remain captured by regressive tendencies arising from their particular social contexts to the extent that they are unable to identify and respond to how those tendencies shape and use them. The issue is not that individual philosophers are sexist, racist, Eurocentric, or overtly committed to the militarism and profiteering which Western sciences so robustly serve these days. Some may well be. However, the point here is rather that mainstream philosophy of science encourages a very weak and thin appreciation among scientists and philosophers of how the conceptual frameworks and the actual projects of sciences and their philosophies have been and are located in historical contexts.[2] Mainstream philosophy of science, whether articulated by philosophers or scientists, often feels like a sleepwalker, wandering around under the spell of a narrative with little connection to the world it seeks to navigate. Whatever its conscious intentions, in such work the sources of truth and of social power amicably cohabit. To the extent that philosophers cannot critically locate their work in real-time social contexts, the work remains part of the problem for politically progressive social tendencies. Of course many philosophers do not evaluate feminism, postcolonialism, and possibly even criticisms of white-supremacy as progressive. But that is a different problem. I refer here to philosophers who do find such tendencies socially progressive, but do not in fact engage with the philosophic implications of their criticisms of the sciences.

Second, the model of philosophy as a mere "handmaid" to the sciences needs a vigorous overhaul. Philosophy of science needs to

---

[2] Especially interesting here are the recent histories of science's regulative ideals. See, for example, Daston and Galison, 2007; Hollinger, 1996; Lloyd, 1984; Mirowski, 2005; Schuster and Yeo, 1986; Shapin, 1994.

move out in front of how sciences are conceptualized, institutionalized, and practiced within the global political economy today. Modest adoption of the "handmaid" role could have been defensible when Western sciences themselves struggled on progressive intellectual, ethical, and political frontiers. It appears ridiculous—at best—today, when the funding and design of scientific projects are so clearly directed primarily by interests in profiteering and militarism.

It is not that no politically progressive motivations direct scientific work; of course they do, for example in health and medical research and environmental sciences. Nor is the claim here that scientists and philosophers are themselves committed to profiteering and militarism. Mostly they are not. Rather, scientific work is usually very expensive, and so the scientific problems that will be able to attract funding will tend to be those of interest to people with lots of money and with access to the parts of nature that are the object of their study. Poor people's knowledge needs in the West or around the globe tend not to be prioritized unless such needs also serve dominant interests. To be sure, such needs can coincide when, for example, the issue is communicable diseases, or the maintenance or production of resources desired by dominant groups (such as safe air, water, and food for the dominant groups). The studious insistence in mainstream philosophy of science on keeping one's intellectual gaze far above the everyday realities of scientific practice in the global political economy, up in the world of abstract formal relations, leaves such philosophy part of the problem for progressive a social movements. If the philosophies of science are to escape being merely philosophies of the kind of knowledge production desired by dominant groups, philosophy needs to develop resources to enable it to escape the tendencies to orbit power which are so prevalent in the sciences themselves. Philosophic projects do not require the massive funding necessary for scientific ones! Philosophy needs to critically evaluate how such power and its philosophical orbiters affect the kind and quality of knowledge so produced.

Third, the production of scientific knowledge in the West entered a new phase with the end of the Cold War in 1989. Its institutional structures shifted, and so did the extent to which it had to respond to various forms of larger community oversight. Already, sociologists and political philosophers are producing accounts of these shifts which either explicitly or implicitly propose very different understandings of the sciences' actual and desirable regulative ideals than those shaped by conventional philosophic understandings. (See, for example,Beck, 1992; 1997; Gibbons et al., 1994; Kleinman and Vallas, 2001; Nowotny, Scott, and Gibbons, 2001; Sassen, 1998.) Either philosophers will abandon the understandings of these ideals which are still dominant in philosophy, or they will become even more archaic than they already are as they continue to work out in even greater detail regulative ideals for which the kind of science to be so regulated has long disappeared.

Finally, philosophies of science could play a useful role in attracting "counter-modernities" into peaceful and fruitful international relations. What are counter-modernities? The phrase is UlrichBeck's (1997). Counter-modernities are social movements or forces which resist the Western modernization processes to which sciences and technologies have always been central. One can think of many examples of such counter-modernities, such as those defended by religious fundamentalisms in the U.S. and in other parts of the world.Beck's account is especially interesting in that he delineates how these are in fact for the most part produced or exacerbated by particular modernization projects themselves. Trying to openly defeat them simply hardens their resistance. Rather, he proposes that progressive forms of modernization need to figure out how to recruit the progressive tendencies within such counter-modernities to projects of the emerging "second modernity" or "reflexive modernity" which lack the flaws of the first, industrial era modernization projects. Philosophies of science can have a powerful role in identifying such progressive strains buried in counter-modernities, and in developing philosophies that can recruit them to scientifically and politically progressive projects. (See Harding, 2008.)

3. How has your work offered original contributions to discussion on science? What does your work reveal that others fail to appreciate? *and* **4. What is the relation between philosophy of science and scientific practice, science policy, or efforts for social justice? Can there be a more productive relation? Is this desirable?**

What have been my original contributions to discussions about science? Before answering such a self-serving question (!), I point out a characteristic of my work shared with many other philosophers. I have always worked in solidarity with one or more progressive social movements. Initially it was thepost-positivist philosophy and social studies of science and technology movement. (See Hackett et al., 2008, for the latest collection of such work.) This was in fact not only an intellectual movement but also social movement very much in the spirit of the challenges to top-down authority characteristic of other powerful social movements of the 1960s in Europe and the U.S. Subsequently I worked in solidarity with the ideals and methodologies of feminist, anti-racist, andpostcolonial global social justice movements. I think that my contributions to philosophy of science have been strongly shaped by historical processes and lively discussions in all of these social movements.

One way in which this collective context is evident in my work is in the many readers I have edited. My first book was such a collection arising out of my dissertation issues (Harding, 1976), and I am now working on my tenth subsequent reader (Harding, forthcoming), this one on issues in the field ofpostcolonial science and technology studies, including feminist issues. These readers have been designed as both scholarly and classroom contributions. As contributions to research and scholarship, they have usually enacted a kind of literature review, showing the conflicts, debates, and tensions within a literature that is scattered across disciplines and that has serious critical implications for mainstream philosophic work. Sometimes they have overtly tried to insert the issues of social movements into mainstream philosophy and/or science studies. As classroom texts they have been designed to enable faculty either to use these essays as the main focus of a course, or as an historical and interdisciplinary background to their own discipline-specific knowledge.

Though the particular form my contributions and projects have taken may be more or less peculiar to me, and more or less earlier or later than those of others working on such projects, like these social movements I see philosophical work in and on the sciences as deeply socially-constituted by its historical moment. At their best, these movements all criticize the "possessive individualism," and its exceptionalist and triumphalist stance that have been regulative ideals for modern work in philosophy and other research and scholarly areas. Thus my work is critical of the ideal that the accomplishments of modern Western sciences and their philosophies owe no significant intellectual debts to the scientific and technological traditions of other societies (the exceptionalist ideal), and of the assumption that "real science," "basic science" and its philosophies has no social/political downside: any bad social effects are the consequence of politics, not of the sciences proper and their philosophies. In my view, as in the view of these social movements, modern Western sciences and their philosophies are deeply implicated in many recent social tragedies and disasters, from environmental destruction to "risk society" to discriminatory politics against various exploited social groups.

Now to the contributions. My work in edited and authored books and many essays has contributed to expanding the domain of epistemology and philosophy of science in three directions. Sciences are necessarily embedded in social worlds just as are any other human activities. In all three ways this work insists on a more robust and realistic notion of "the social" in the contexts which create the "social construction of knowledge." In the first case, we included gender relations in the historical eras with which sciences have an integrity, as Thomas S. Kuhn (1970) put the point. Thus feminism is not only a moral and political issue, but also one for epistemological and philosophy of science focus. The second expansion brings postcolonial science and technology studies, and especially its histories of sciences North and South, into mainstream epistemology and philosophy of science. (See Selin, 1997; Harding, 1998.) Indeed this work tries to replace the now empirically and theoretically unsupportable histories of science in which mainstream philosophies ground their underdeveloped discussions of regulative ideals with the new feminist and postcolonial histories of science.

A third expansion of the domain of mainstream philosophy of science, visible in my most recently published book (Harding, 2008), tries to set philosophical issues in the context of modernity studies (e.g., Eisenstadt, 2000). It gives the modernity/tradition binary the same kind of critical interrogation that postmodernists have given to modernity. It is important to recognize that this kind of interrogation of both the concept of tradition and its persistent contrast with modernity was not possible until postcolonial studies had begun to argue for the existence of multiple modernities, each with its own scientific and technological traditions. Of course in modern sciences, their epistemologies and philosophies have been constituted through the modernity-vs-tradition conceptual framework, so it has required the critical historicization of that contrast to enable an empirically and theoretically more defensible understanding of modernity. My work here asks what modernities designed *for* women and the West's other others would look like. What philosophies of science would be appropriate for such modernities?

In all three cases, I have intended to open up new discussions of relations between everyday experience and the production of scientific knowledge—an issue prefigured in philosophy at the middle of the last century by Quine and Kuhn, among others. Two related methodological tools have facilitated this work: standpoint theory and a regulative ideal of "strong objectivity." Standpoint theory has an older Marxian legacy, but as Frederic Jameson (2004) points out, it was only in feminist work of the 1970s and early 1980s that its resources were identified again and transformed for creating the possibility of new and useful "polemics about the relations between everyday experiences and scientific knowledge." Standpoint theory was initially developed in the sociology of knowledge by Dorothy Smith (1987; 1990), and a few years later and independently in the work of political philosophers Nancy Hartsock (1983) and Alison Jaggar (1983), as well as sociologist of science Hilary Rose (1983), and in my own work.

While there are differences between how the issues are identified and developed—differences created by attention to distinctive disciplinary discourses, among others—all these theorists argue that the research disciplines are complicitous with power, working up daily life into the kinds of categories and the causal relations between them that enable the few to govern the many. Sciences that are *for women* or other marginalized groups must start off thinking from the daily lives of people in such groups to discover otherwise hidden realities about everyday life, how dominant social institutions function, and, most importantly, the relations between everyday life and the principles and practices of such institutions. For example, starting off thinking about medical and health practices from women's daily lives, instead of from the interests of institutions run by health profiteers and medical experts, has resulted in many successful challenges to the principles and practices of such institutions. It takes "strong objectivity," which overtly values disvalued everyday experience, to reveal such realities, as I have argued. Standpoint theory is useful not only as an epistemology and philosophy of scientific research, but also as a methodology: a directive for how to do good research.

In the hands of critics of the modernity vs. tradition binary, standpoint methodologies can reveal just whose social progress modernization theories and practices in fact benefit, and which social groups have borne a disproportionate share of their costs. Modern sciences and technologies function as one-way time machines in modernization theories and practices, transporting into the "public sphere" individuals (usually only men) and such universally created social institutions as those that govern economies, political systems, education/science, and moral and religious values. Left behind is a now vastly weakened "private sphere" of household, family, kin, and tribe—namely the women and their worlds which in modernity are exploited and abandoned by the public sphere's regulative ideal of "social progress." This kind of critical account of Western modernity's way of constituting its relation to its other, tradition, reveals the powerful co-constitution of "the feminine" and "the savage" or "the primitive" as the other, the threatening enemy to be overcome, if modernity's social progress is to flourish. Modernity is haunted by specters of the feminine and the savage, I argue in this work. It is only when the projects of feminism, anti-racism andpostcolonialism are mutually supportive that such specters can be "exorcised" from the heart and soul of the West's regulative ideals of social progress.

5. Where do you see the field of philosophy of science to be headed? What are the prospects for progress regarding the issues you take to be most important?

What are the prospects for new directions in the philosophy of science? On the one hand, current regulative ideals in the philosophy of science mimic the modernization rhetorics of the Nineteenth Century industrial revolution and the 1950s European/American economic, political, and technological expansion. Scientific practices and their philosophic regulative ideals in the West today are thus both intellectually and politically regressive. Insofar as philosophy of science work is contained by such modern Western conceptions of social progress, it remains part of the problem and is widely recognized as such around the globe. Such philosophies of modern Western sciences provide resources for "the conceptual practices of power," (Smith, 1990).

On the other hand, the groundwork for resourceful and energetic philosophies of science are emerging from various other disciplines—history and sociology of science, especially. And from social justice movements. Thus this is an exciting historical moment to identify the margins to occupy and horizons to cross in search of sciences and their philosophies that are for the vast majority of the world's populations, rather than for delivering yet more resources to the already over-privileged few. One can be sure of having intellectually and politically lively and welcoming company on such quests.

References

Beck, U. *Risk Society: Towards a New Modernity*. London: Sage Publishing Co., 1992.

Beck, U. *The Reinvention of Politics: Rethinking Modernity in the Global Social Order*. Cambridge, U.K.: Polity Press, 1997.

Boston Women's Health Collective. *Our Bodies, Ourselves*. Boston: New England Free Press, 1970.

Daston, L., and P. Galison. *Objectivity*. New York: Zone Books, 2007.

Eisenstadt, S. N. "Multiple Modernities." *Daedalus*. 129(1): 1-30, 2000.

Feyerabend, P. *Against Method*. London: New Left, 1975.

Gibbons, M. et al. *The New Production of Knowledge: The Dynamics of Science and Research in Contemporary Societies*. Thousand Oaks: Sage Publications, 1994.

Hackett, E. J., et al. (eds.) *The Handbook of Science and Technology Studies. 3rd Edition*. Cambridge: MIT Press, 2008.

Hartsock, N. "The Feminist Standpoint." In S. Harding and M. Hintikka (eds.), *Discovering Reality: Feminist Perspectives on Epistemology, Metaphysics, Methodology, and Philosophy of Science*. Dordrecht: Reidel, 1983.

Hollinger, D. *Science, Jews, and Secular Culture*. Princeton: Princeton University Press, 1996.

Jaggar, A. "Chapter 11: Feminist Epistemologies" In A. Jaggar, *Feminist Politics and Human Nature*. Totowa, N.J.: Roman and Allenheld, 1983; reprinted in Harding (ed.), 2004.

Jameson, F. "'History and Class Consciousness' as an Unfinished Project." In S. Harding (ed.), *The Feminist Standpoint Theory Reader*. New York: Routledge, 2004.

Kleinman, D. L., and S. P. Vallas. "Science, Capitalism, and the Rise of the 'Knowledge Worker': The Changing Structure of Knowledge Production in the United States." *Theory and Society*. 30: 451-492, 2001.

Kuhn, T. S. *The Structure of Scientific Revolutions, 2nd Edition*. Chicago: University of Chicago Press, 1970.

Lloyd, G. *The Man of Reason: "Male" and "Female" in Western Philosophy*. Minneapolis: University of Minnesota Press, 1984.

Mirowski, P. "How Positivism Made a Pact with the Postwar Social Sciences in the United States." In *The Politics of Method in the Human Sciences: Positivism and its Epistemological Others*. Durham: Duke University Press, 2005.

Nowotny, H., P. Scott, and M. Gibbons. *Re-Thinking Science: Knowledge and the Public in an Age of Uncertainty*. Cambridge, U.K.: Polity Press, 2001.

Quine, W. V. O. *Word and Object*. Cambridge: MIT Press, 1960.

Rose, H. "Hand, Brain and Heart: A Feminist Epistemology for the Natural Sciences." *Signs*. 9: 73-90, 1983.

Sassen, S. *Globalization and its Discontents*. New York: The New Press, 1998.

Schuster, J. A., and R. R. Yeo (eds.) *The Politics and Rhetoric of Scientific Method: Historical Studies*. Dordrecht: D. Reidel, 1986.

Selin, H. (ed.) *Encyclopedia of the History of Science, Technology, and Medicine in Non-Western Cultures*. Dordrecht: Kluwer, 1997.

Shapin, S. *A Social History of Truth*. Chicago: University of Chicago Press, 1994.

Smith, D. *The Everyday World as Problematic: A Sociology for Women*. Toronto: University of Toronto Press, 1987.

Smith, D. *The Conceptual Practices of Power*. Toronto: University of Toronto Press, 1990.

Tobach, E. and B. Rosoff. (eds.) *Genes and Gender, Vols. 1-4*. New York: Gordian Press, 1978-1984.

Selected Bibliography

Can Theories Be Refuted? Essays on the Duhem-Quine Thesis. Editor. Dordrecht: Reidel, 1976.

The Science Question in Feminism. Ithaca: Cornell University Press, 1986.

Feminism and Methodology: Social Science Issues. Editor. Bloomington: Indiana University Press, 1987.

8. Sandra Harding

Whose Science? Whose Knowledge? Thinking From Women's Lives. Ithaca: Cornell University Press, 1991.

The 'Racial' Economy of Science: Toward a Democratic Future. Editor. Bloomington: Indiana University Press, 1993.

Is Science Multicultural?Postcolonialisms, Feminisms, and Epistemologies. Bloomington: Indiana University Press, 1998.

Decentering the Center: Philosophy for a Multicultural, Postcolonial, and Feminist World. Co-edited with U. Narayan. Bloomington: Indiana University Press, 2000.

Discovering Reality: Feminist Perspectives on Epistemology, Methodology, Metaphysics, and Philosophy of Science. Co-edited with M. Hintikka. Dordrecht: Kluwer Publishing Co, 1983; Second Edition (Twentieth Anniversary Edition), 2003.

Science and Other Cultures: Issues in Philosophies of Science and Technology. Co-edited with R. Figueroa. New York: Routledge, 2003.

The Feminist Standpoint Theory Reader: Intellectual and Political Controversies. Editor. New York: Routledge, 2004.

Science and Social Inequality: Feminist andPostcolonial Issues. Chicago: University of Illinois Press, 2006.

Sciences From Below: Feminisms,Postcolonialities, and Modernities. Durham: Duke University Press, 2008.

"Interrogating the Modernity vs. Tradition Contrast: Whose Science and Technology for Whose Social Progress?" In H. Grasswick (ed.), *Feminist Epistemology.* Dordrecht: Kluwer, 2009.

ThePostcolonial Science and Technology AfterPostcolonial Theory: A Reader. Editor. forthcoming.

9

Don Ihde

Distinguished Professor of Philosophy
Stony Brook University, USA
Technoscience Research Group, Department of Philosophy

1. How were you initially drawn to philosophical issues regarding science?

It is said Leonardo da Vinci's Renaissance sized library had 166 books—just about the size of the small library of my one-room elementary school in Kansas. It was the set of *Encyclopedia Britannica* which drew my strongest attention and I avidly read article after article in the generous spare time available in such a school. Dinosaurs, evolution (not under any fundamentalist clouds in those days), astronomy and other science topics became my passion. Later, as an undergraduate at the University of Kansas, physical anthropology, human origins, vertebrate paleontology continued to be fascinating but now also including my minor in philosophy: philosophy of language, theory of meaning and epistemology. Much later, but still early in my philosophical career, I became interested first in the philosophy of technology with *Technics and Praxis* (1979) my first foray and with an opening relating to science's embodiment through instruments, and then a more distinct turn to philosophy of science with *Instrumental Realism: The relation between the philosophy of science and philosophy of technology* (1991). By then, I understood myself as a 'phenomenological realist' and this interest in science in its instrumental practice continues to the present.

2. What, in your view, are the most interesting, important, or pressing problems in contemporary philosophy of science?

It appears to me that what could be called the *image* of science has been drastically altered since the mid-20th century into the 21st.

First, from the anti-positivists, later from the new sociologies of science and the emergence of interdisciplinary science studies, I would say that the understanding of how science produces robust knowledge has grown both broader and more complicated. The earlier, and I would say 'thin' notions of science as a sort of theory-producing machine, have been replaced with the recognition that science as a human activity is also *situated*—in culture, with practices including social dimensions, relations to the wider political and normative aspects of contemporary life, and above all, including both a material culture and material laboratory and instrumental activities. Moreover, as results from the "culture wars" and "science wars" that have been propagated in more popular media and played out in governmental legislatures and corporate industries, questions of trust in science, the changing roles for objectivity, and freedom for questions of research emerge as highly important today.

Much of this relates to the radical changes which have occurred within modern science, and to late modern, even postmodern science. Big Science, with corporate-like structures, mega-instruments and mega-interdisciplinary teams, are no longer the early War and post-War exceptions, but are the norm. Science, and the philosophy of science, must take this transformation into account. Here the now several decade old debates concerning postmodernity, relativism, and the social dimensions of science, although not resolved, have had impact. The more popular expressions of these "wars" were often distortions, and in my own case I consider myself to be sympathetic to some aspects of postmodernism insofar as it stresses situated knowledge, an embodied and culturally embedded understanding of knowledge, and the development of multiple perspectives (in my case through multiple instrumental perspectives). But such a more concrete and even 'existential' understanding of knowledge is not necessarily a *relativism*. I have usually distinguished between *relativism* and *relativity* and consider my own epistemology and ontology to be *relativistic* in the sense that there is an interaction between knower and known, but in which interaction can be described and accounted for.

3. How has your work offered original contributions to discussion on science? What does your work reveal that others fail to appreciate?

My own work, I hope, has offered a few original or at least once minoritarian contributions to contemporary philosophy of science. First, as noted in the first question, my turn to the role of instruments in science was an early one—it preceded the work of Ian Hacking, Robert Ackermann, and Peter Galison, all of whom I consider important contributors to this focus. In my case, I emphasized the role of instrumental *materiality*, the necessary relations to human *embodiment*, and the phenomenological-analytic attention to the way in which instruments are *non-neutral variations* which expand and transform human perception and knowledge. I have argued that the actional character of the use of instruments in scientific practice is a type of materialized phenomenological variation, an instrumental variation, which produces new knowledge in science's investigations. But even more, instruments suggest developmental trajectories which often lead or open the way to new domains and levels of knowledge. By following the macro and micro developments—for example in imaging technologies—the entire universe of science is expanded.

The second contribution is, I would hold, an understanding of the critical process of science as a *hermeneutic* endeavor, a critical interpretive activity. In *Expanding Hermeneutics: Visualism in Science* (1998), I try to show, not only that science in most of its practice has produced a kind of visualist sub-culture, but has developed a highly sophisticated *visual hermeneutics* which, unlike traditional language-embedded hermeneutics, is more perceptual and image-oriented. In this case I would recognize both Peter Galison with his understanding of the traditions of image and logic in physics, and Bruno Latour with his understanding of laboratory devices as visualization displays as complementary thinkers sensitive to this emphasis.

The third emphasis of mine has been to try to show how, implicitly if not always explicitly, science must take account of human *embodiment*. For example, in the radical new image technologies which image far beyond the early modern limits of the optical range, now reaching from gamma ray to radio wave radiation, images—to be read or interpreted—must be *translated* back into humanly perceivable images. The perceptual gestalt qualities embodied humans have, can recognize patterns precisely when such translation processes are successful. Our specific bodily capacities are thus part of the entire constructive process of making scientific knowledge.

4. What is the relation between philosophy of science and scientific practice, science policy, or efforts for social

justice? Can there be a more productive relation? Is this desirable?

Another important shift in science and the philosophy of science, in which the issues of policy, social justice—and I would add multiculturality and gender issues—become forefronted, lies in the shift of what sciences are dominant and to which dimensions of social-political life these relate. Some historians of science have argued that the cancellation of the Super Collider was a result of the end of the Cold War. Much science, particularly physics and many engineering sciences, were clearly highly embedded in military supported funding, and while sub-atomic processes might seem very far distant to potential weaponry, the tacit beliefs of politicians may have seen the relation differently. Of course the indirect beneficiaries then became, not the physical, but the biological sciences, which now themselves become Big Science with the genome and now genetic and biotechnological sciences. These shifts are, of course, complicated and engage ever new and different groupings of society.

I am reminded here of work done by feminist philosophers and science studies thinkers. Whatever opposition and criticism early feminist science critics may have undergone, it can now be seen that in contemporary notions of evolutionary reproduction strategies, the early revolutions in primatology and the later revolutions in DNA mating marking which changed the paradigm of animal reproduction, were in fact often led by women whose questions were different from those of their male predecessors. A shift from an almost exclusive focus upon impregnation strategies to one which includes infant protection and preservation strategies was one factor in this shift. Here the calls for more democratic and pluralistic representation within science itself remain of high importance and I am happy to consider my own thinking as having been helped by such science critics as Donna Haraway and Sandra Harding. But we have a long way to go.

5. Where do you see the field of philosophy of science to be headed? What are the prospects for progress regarding the issues you take to be most important?

I believe that the trends outlined above will continue to be important to the philosophy of science. I regard the more modest and more situated knowledge approach to science to be a healthy one. Fallibilism can serve philosophy of science to good purposes. And, I would also return here to the important role of science's material culture and its instruments. There has been, since the 20th century, what I call a 'second scientific revolution' related to instrumentation and imaging processes. Until the accidental discovery of radio astronomy early in the last century, all astronomy had been limited to optical light, but today different instruments detect and image the microwave spectrum from gamma to radio waves. Then, with the addition of not only the computational power of computers, but now the digital data-to-image conversions and the development of highly complicated modeling processes, 21st century science looks very different not only from its 17th century beginnings, but even from its late modern peak in the 19th century. Today's science is also interdisciplinary, from physics to medicine, and I wonder if philosophy, too, might not benefit as well from becoming interdisciplinary and project oriented.

Selected Bibliography

Authored Books

Hermeneutic Phenomenology: The Philosophy of Paul Ricoeur. Evanston: Northwestern University Press, 1971.

Sense and Significance. New York: Humanities Press, 1973.

Listening and Voice: A Phenomenology of Sound. Athens: Ohio University Press, 1976; second edition Albany: SUNY Press, 2007.

Experimental Phenomenology. New York: G. P. Putnam's Sons, 1977; Albany: SUNY Press, 1986.

Technics and Praxis. Dordrecht: Ridel, 1979.

Consequences of Phenomenology. Albany: SUNY Press, 1986.

(monograph) *On Nonfoundational Phenomenology.* Publikationer fran institutionen for pedegogik, Fenomenografiska notiser 3. Goteborgs, 1986.

Technology and the Lifeworld. Bloomington: Indiana University Press, 1990.

Instrumental Realism: The Interface Between Philosophy of Technology and Philosophy of Science. Bloomington: Indiana University Press, 1991.

Philosophy of Technology: An Introduction. New York: Paragon House, 1993.

Postphenomenology: Essays in the Postmodern Context. Evanston: Northwestern University Press, 1993.

Expanding Hermeneutics: Visualism in Science. Evanston: Northwestern University Press, 1998.

Bodies in Technology, Electronic Mediations Series, vol. V. Minneapolis: University of Minnesota Press, 2002.

Ironic Technics. New York: Automatic Press/VIP, 2008.

Postphenomenology and Technoscience: The Peking University Lectures. Albany: SUNY Press, 2009.

Edited Books

Phenomenology and Existentialism. Co-edited with R. M. Zaner. New York: Capricorn Books, 1973.

Paul Ricoeur, The Conflict of Interpretations. Evanston: Northwestern University Press, 1974.

Dialogues in Phenomenology. Selected Studies in Phenomenology and Existential Philosophy, vol. V. Co-edited with R. M. Zaner. The Hague: Martinus Nijhoff, 1977.

Interdisciplinary Phenomenology. Selected Studies in Phenomenology and Existential Philosophy, vol. VI. Co-edited with R. M. Zaner. The Hague: Martinus Nijhoff, 1977.

Hermeneutics and Deconstruction. Selected Studies in Phenomenology and Existential Philosophy, vol. IX. Co-edited with H. Silverman. Albany: SUNY Press, 1985.

Descriptions. Selected Studies in Phenomenology and Existential Philosophy, vol. X. Co-edited with H. Silverman. Albany: SUNY Press, 1985.

Chasing Technoscience: Matrix for Materiality. Co-edited with E. Selinger. Bloomington: Indiana University Press, 2003.

Special Issue on Postphenomenological Research, *Human Studies.* 31(1), 2008.

Recent Genre Review Articles

"Technoscience and the 'Other' Continental Philosophy." *Continental Philosophy Review.* 33: 59-74, 2000.

"Has the Philosophy of Technology Arrived? A State of the Art Review." *Philosophy of Science.* 71(1): 117-131, 2004.

10

Sheila Jasanoff

Pforzheimer Professor of Science and Technology Studies
Harvard University, John F. Kennedy School of Government, USA

An STS Scholar Reflects on Philosophy

1. How were you initially drawn to philosophical issues regarding science?

Not all at once, nor easily. Philosophical questions were far from my thoughts when I first entered the field of science and technology studies (STS)[1] at Cornell University in the late 1970s. The field's history, my own intellectual biography, and the institutional circumstances at Cornell all militated against a deep and positive engagement with philosophy. That I later found it essential to relate my work to aspects of the philosophy of science says much about the changing preoccupations of STS and about the nature of interdisciplinarity.

Like most other boundary-crossing intellectual activities, science studies evolved by different routes in different times and places. European developments were substantially more entwined with philosophical concerns from early on than were parallel developments in the United States. For me, a convenient way to explain my interest in the philosophy of science is to recall these separate histories and what they have meant for my work. I begin with the struggle over the meaning of "STS" itself, a struggle in which I found myself unexpectedly caught up in the 1980s, and which captures to some degree the tensions I have felt in shaping my encounters with philosophy.

---

[1] For convenience, I use the terms "science and technology studies," abbreviated STS, and "science studies" interchangeably throughout this text. Far from shortchanging technology, I wish to emphasize that science studies is incomplete without the full inclusion and incorporation of technology studies as integral to the field.

At Cornell, as at MIT and other U.S. universities in the 1970s, STS stood for "science, technology, and society." Concerns about the intersection of science and technology on the one hand and society on the other bubbled up out of the cauldron of discontent and invention that was the "1960s." Whichever way one turned, the products of science and technology seemed to be allied with violence and destruction: Napalm and Agent Orange in Vietnam; nuclear power plants endangering fragile aqueous ecosystems and threatening worldwide atomic fallout; chemical pesticides polluting the environment and silencing birdsong[2]; the beginnings of artificial intelligence and biochemical control of the mind; and, with heightening intensity from the launch of Sputnik in 1957, the Cold War specter of satellite-borne surveillance and global warfare. Naively, unanalytically, but passionately, many in America felt that these evils had been, or were about to be, inflicted on an unprepared world with little criticism and no accountability. STS programs sprang up at many universities to provide a space where these concerns could be aired and articulated. At Cornell, a course entitled "The Impact and Control of Technological Change," designed in 1971 with funding from the National Science Foundation, carried this vision of STS into the classroom (Nelkin, Hershey, and Mueller, 1974). It was still being taught when I arrived at Cornell in 1978. By black-boxing the inner workings of technology, the course ratified, indeed some would say enacted, the split between the normative and the metaphysical and epistemological that was, for long, a defining feature of STS in America.

In European academic circles, the black-boxing ran to some extent the other way, again under the rubric of STS, but now understood as "science and technology studies." Here, in contrast to the American scene, the conversation between STS and the philosophy of science was absolutely central, as sociologists challenged philosophy's historical monopoly over questions of truth and facticity. The strong programme in the sociology of scientific knowledge (SSK), most prominently associated with the work of David Bloor (1976), famously viewed scientific claims as social constructs that could be subjected to painstaking empirical investigation; at the same time, SSK left largely unanalyzed such basic social categories as "interest." The French school of actor-network

---

[2] The birth of the modern environmental movement in the United States is often dated to the publication of Rachel Carson, *Silent Spring* (1962).

theory (ANT) did not fall into the trap of social determinism, but in their primary concern with non-human, material entities, and the place of objects in social life, leading ANT scholars such as Michel Callon and Bruno Latour side-stepped the concerns for equity and justice, and the control of power and violence, that had mobilized American support for STS in the 1960s and 1970s.[3]

I myself wandered into these unsettled debates about the definition, purposes and methods of STS almost by accident. As a trained lawyer with two years of environmental practice behind me, I found myself from 1978 onward carving out an academic niche in Cornell's tiny Program on Science, Technology and Society. Templates for what I should be doing were almost nonexistent. Dorothy Nelkin's (1992) work on technical controversies was the most readily available model, but although prodigious in number and wide-ranging in subject matter, the case studies she wrote and commissioned did little to spark a deeper interest in the nature of knowledge or the reality of things. The explanation for controversy in all the cases seemed to be the same: science is often uncertain, and in contested domains disparate actors construe scientific evidence to suit their social interests. Not only were the interests taken for granted, but science itself was treated as essentially homogeneous, with little attention given to the detailed work of generating different kinds of claims and counterclaims in different areas of science. Even moral issues were largely elided in American STS work of that generation, since many believed that the knowledge of those in power was always biased against the lay public's interests, and the researcher's duty was to expose the biases of power through a selective and necessarily asymmetrical unmasking of the state's expertise. Needless to say, this stance left little space for thinking about the normativity of scientific observations—or, put differently, about better and worse ways of representing reality, or what might be at stake in making such normative judgments.

My first forays into STS research occurred, somewhat unusually, under the heading of comparative politics. Usually conducted by political scientists, such research tries to understand why nation states so often choose to pursue similar social objectives, such as environmental protection or economic growth, with markedly different strategies. A project on chemical regulation in Europe and the United States showed that different national authorities

---

[3] See particularly Michel Callon, 1986.

frequently assess the risks of the same cancer-causing substances differently, although such bodies are equally concerned about public health and safety, respond to very similar legal mandates, and rely on the same scientific studies (Brickman, Jasanoff, and Ilgen, 1985). My attempts to understand this anomaly led slowly but surely to a productive encounter with what was then still known as "social constructivism" (the adjective "social" was later dropped to a large extent). It seemed clear to me that features of social organization and behavior importantly constrain the interpretations that public authorities are able, or willing, to place on disputed or uncertain knowledge claims. In my first comparative explorations, I took those organizational features as mostly fixed and static in given national contexts: for example, my co-authors and I attributed some of the most significant variations in risk assessment between the United States and Europe to the greater transparency and adversarial, legalistic format of U.S. regulatory processes.

Over time, however, I came to find these structural explanations for variations in public reasoning and fact-finding less and less compelling. For one thing, they could not easily account for significant shifts in policy discourse or consciousness, as for example the move in Europe from a risk-based to a precaution-based approach to coping with environmental hazards, or in the United States from an end-of-pipe approach to controlling pollutants to one that tries to prevent the very phenomenon of pollution. For another, the explanation that "scientific knowledge advanced" lost its seductive appeal. It seemed to offer no more than a descriptive restatement of what had happened; it only opened up the secondary question of how new knowledge had gained ground over prior conventional wisdom. Different national authorities, moreover, seemed to take up new knowledge claims with markedly different degrees of receptivity. Also, as I read more widely in the science studies literature, it became increasingly less satisfying to account for social and political power without including some analysis of public knowledge and public beliefs. In this way, epistemology found me when I was hardly looking for it.

I did not get much help in thinking through my problems from colleagues in the philosophy of science—possibly because the questions germinating in my mind were as yet half-formed and were framed in ways unfamiliar to philosophers. The few Cornell philosophers who took an interest in STS in those days were preoccupied with the seemingly never-ending battle between realism and relativism, the latter quintessentially represented for them by the Edinburgh school of SSK and by some of Latour's more provocative writings. That debate, they were quite convinced, had been handily won by the realists. The "science wars" of the 1990s gave considerable aid and comfort to both philosophical and natural realists,[4] although, curiously and fortunately, they in no way disrupted the cordial relations between Cornell's science faculty and the relatively young Department of Science and Technology Studies, formed in 1991. Epistemologically, however, my philosophy colleagues seemed to consider the issues closed. As one distinguished philosopher told me sometime in the mid-1990s, the only connection left between philosophy and STS was in the domain of ethics and moral philosophy.

In that domain, moreover, the specific institutional history of STS at Cornell made interactions difficult. For many trained moral philosophers, new areas in science and technology, such as the issues connected with genetics and biotechnology, simply provided new ground for the application of old principles. As if to illustrate the point, one disgruntled student complained to me that he had taken both bioethics and environmental ethics in the Cornell philosophy department and found no philosophical differences worth recounting: the courses were the same; only the illustrations varied. By contrast, the STS department approached bioethics from the standpoint of practice and asked how people involved in therapeutic processes, whether as patients or caregivers, perceive and deal with moral choices. Of necessity, this latter approach has to be grounded in the specificities of biomedicine. A course in medical ethics *could* not look the same as a course in environmental ethics from the actor-centered perspective of STS, since each would induce its questions and principles from a different domain of practice. Moreover, STS denies the proposition that "real" ethical principles exist in a domain of pure analytic ab-

---

[4] The "science wars," conventionally described as a battle between scientific realists and postmodern relativists, began with the publication of Paul R. Gross and Norman Levitt, *Higher Superstition: The Academic Left and Its Quarrels with Science* (1994). The most public event in the controversy was the "Sokal hoax" of 1996. The physicist Alan Sokal contributed an article to the literary journal *Social Text* claiming that quantum physics supports postmodern ideas about science. Later, in a different publication, Sokal revealed that his article was scientific nonsense, meant to test the credibility of *Social Text*'s editors, who had published it uncritically because it was consistent with their ideological biases.

straction, and that these are distorted or misunderstood through the phenomenology of popular perception. Rather, for most STS scholars, principles only exist by virtue of their practical enactment in everyday interactions. Almost predictably, then, control over the bioethics curriculum at Cornell became for a time a site of turf warfare between philosophy and STS, but none of us had the meta-language or the academic incentives to work through the underlying intellectual problems with our colleagues in the other discipline.

Unsurprisingly perhaps, it took a painful transition to Harvard in 1998, and a period of serious reflection on the achievements of STS at Cornell, to begin clarifying my understanding of intersections between my work and philosophical issues, particularly those of interest to the philosophy of science. Below, I address the nature and consequences of that gradual enlightenment.

2. What, in your view, are the most interesting, important, or pressing problems in contemporary philosophy of science?

I came to the philosophy of science by back roads, through the double gates of law and sociology of science, and my starting points for asking philosophical questions accordingly were issues of primary importance in the social sciences. That journey shaped my perception of what is important for philosophers to be asking. Increasingly, my own questions began to center on the problem of public reason. Why do modern societies accept certain forms of reasoning as better or more persuasive than others in legal and policy environments? What is the role of culture in shaping a polity's collective sense of reason? How do material objects and artifacts participate in the construction of reason, especially in legal settings broadly defined? Are there principles of reasoning that cannot be relativized—that is, principles that must be accepted as universal rather than as contingent on specific contexts? And, following through to the pragmatic work of legal and political decisionmaking, how should actors in those fields choose between competing reasons when confronted by alternative arguments and models of rationality?

These are not simply theoretical questions; they sit at the heart of today's most salient social and political controversies. Many S&T-related conflicts that I have studied closely over the years lead back in the end to tightly integrated cognitive-normative packages that are rooted in highly specific histories and cultures of government. Thus, it is difficult to separate the predilection for a certain form of objectivity in the United States—based on impersonal judgment and, where possible, mathematical logic— from a political process that renders institutions both exceptionally transparent and exceptionally vulnerable to legal challenge (Brickmann, Jasanoff, and Ilgen, 1985; Porter, 1995). Or, to take another example, the kinds of proofs demanded for the safety of genetically modified crops in Britain and the United States in the late twentieth century seem to reflect particular institutional relations between science and politics in those two countries. Science, in the U.S. context, is often regarded as directly "speaking truth to power"; in Britain, by contrast, expertise is strongly tied to notions of experience and judgment, and is regarded as the bridge between knowledge and power. These disparate understandings are connected in turn to public demands for different forms of epistemic and political reason-giving in the two countries (Jasanoff,

2005). Such patterned variations in the practices of scientific and technical reasoning demonstrate that culture conditions the basic meanings of truth, objectivity, evidence, proof, and facts in public forums.

Examples such as these suggest that some of the most interesting and urgent philosophical issues centering on science today are far removed from science as practiced in laboratories or other traditional scientific and technological workplaces. In brief, as scientific practice has changed—becoming more interdisciplinary, more socially complex, more reliant on instruments and models, more driven by ethical and political concerns, and more accountable to varied publics—it has become harder to define what constitutes the scientific method, the demarcation between good and bad science, or scientific progress. Yet, the need to discriminate between valid and invalid knowledge remains constant, and indeed, if anything, more pervasive and consequential than a generation or two ago. The shift to what has been called Mode 2 of scientific production appears to call for new philosophical questions, or at the very least for old questions to be reformulated in new language (Gibbons, Nowotny, and Scott, 1994).

Clearly, new issues have arisen around the definition of science itself. How, in any complex, modern society, are some claims recognized as legitimately scientific and others not? Labeled the "demarcation problem," that question has a respectable philosophical lineage, for example, in attempts to demarcate science from religion and true from false theories. Sociologists have addressed the same problem under the heading of "boundary work," calling attention to the varied strategies by which communities of scientists and experts create and maintain divisions between disparate technical disciplines and, more broadly, between science and other forms of life (Gieryn, 1999). Complicating these accounts, we now know that not only scientists but society as a whole is constantly engaged in the special forms of boundary work that give rise to demarcations within and around science. Science, put in terms that anthropologists often use, is both an actor's category and an analyst's category; both actors and analysts, moreover, have escaped the confines of academic disciplines and are scattered throughout society, defining and deploying what they regard as science according to their own needs and understandings. In these circumstances, it is harder to explain in philosophical terms why, for example, "climate skepticism" or "intelligent design" cannot be considered genuinely scientific. Foundationalist claims no longer serve as a good enough resource to define what should count as science in society.

Other questions revolve around the nature of human-made artifacts or entities and how to think about them philosophically. We are in a period of rapid change in which accepted classifications of the natural world are in disarray and new entities are being created or acknowledged with bewildering speed. How should we think about these novelties, especially when they seem to cross ancient, almost sacred lines, such as those between species, between human and animal, between objects and persons, and between nature and culture? Of particular urgency are questions about the nature of the human, a category that has come under siege in two ways: because many disciplines are competing for the right to define human nature; and because new technoscientific entities and practices are blurring or eroding the boundaries between the human, the animal, and the machine. Age-old concerns about determinism, for example, have received new impetus from brain imaging and neuroscience, biopharmacology, genetics, experimental economics, and social psychology; at the same time, the production of human-machine cyborgs and human-animal chimeras denies the very possibility of firm material demarcations of the human body from other kinds of entities. Without renewed foundational thinking, these developments bode ill for societal efforts to secure human dignity, preserve autonomy, and achieve ethical agreement among diverse political communities.

Our capacity for kind-making, moreover, does not stop with bringing new, ontologically troublesome material objects into existence. Scientific capabilities are also continually engaged in reorganizing and reordering the conceptual world, making new demarcations, shaping new ontologies, and rejecting ones that no longer carry sufficient meaning or authority. That we do this is not new in itself. Nelson Goodman (1978) called attention a generation ago to our penchant for worldmaking, or creating new versions of reality with stuff already lying around. What I find myself wondering about, though, is how we decide whether we are in the right world at all when each version is freighted not only with theory (a standard concern in the philosophy of science) but with taken-for-granted cultural assumptions that are difficult to tease out, let alone to translate across boundaries.

The entities that abound in the legal and political worlds that I investigate seem at times to defy classification. For example, the distinction between "natural" and "social" kinds breaks down in the face of such constructs as "fine particulate matter" or "endangered species." Such terms seem to refer to things in the natural world that can be grouped together, thus functioning as kinds, but their primary function is to trigger new forms of human behavior. In that sense, these classifications are profoundly interactive (Hacking, 1999): they exist to serve human needs, and they motivate human societies to act in certain ways, for example, by redesigning polluting utilities to keep them from emitting fine particulates or adopting new land use patterns to protect threatened species. What is at stake in producing such entities is not solely the metaphysical question whether they are the right things with which to represent our world, but also the associated normative question whether it is right for us to live in the worlds we populate with these particular things. For me, then, a question that increasingly sums up many of my personal intellectual troubles in STS is this: in law, in regulation, in public policy, how can we judge whether it is the right conceptual and material world that we are inhabiting? How does getting the world right from the standpoint of truth-seeking relate to making the right world from the standpoint of justice?

3. How has your work offered original contributions to discussion on science? What does your work reveal that others fail to appreciate?

The shortest way to characterize my contribution is that it seeks to socialize epistemology. In that respect, my work is coherent with that of many sociologists to science who have also tried to show that philosophical questions, and their answers, have social underpinnings. What makes my work distinctive is that it pursues the sociology of knowledge into the realm of political theory in ways that are not appreciated by all readers of my oeuvre, least of all those in political philosophy.

Since much of my work rests on thick empirical descriptions, it is sometimes difficult for casual readers to step away from the substance and discern the larger themes that are explored in a body of writing that spans thirty years, four countries, and at least three major topical areas: environment, biotechnology, and science policy. The fact that my interests can also be described under policy-relevant or "applied" headings, such as risk or sustainability, may also work against conversations of a more theoretical nature. In disciplinary terms, too, my work does not fit neatly within recognized paradigmatic boxes but cuts across law, political science, and historical and sociological studies of science and technology. There is, however, a major analytic theme that connects all of my disparate strands, namely the construction and deployment of *public reason*: how is it that powerful social actors of all kinds—courts, regulatory agencies, expert bodies, and citizen groups, among others—persuade citizens that their arguments make sense, hold together, are reasonable, and are therefore entitled to respect?

In playing around with this question for more years than I like to count, I have made several theoretical observations that are worth disentangling for philosophical audiences. The first has to do with the nature of *objectivity* in public domains. Together with a considerable body of historians and philosophers of science, I argue that objectivity is contingent on time and place, that its ingredients are culturally specific, and that one era's objectivity can be another era's bias or outright mistake.[5] Unlike many of my STS colleagues, however, I treat objectivity as a cultural resource that is as important to the conduct of law and politics as it is of science. Indeed, the peculiar objectivity of science, I argue, is threaded through with deep underlying ideas about how to get to the place where a public claim will be considered "objective," that is, untainted by bias or prejudice, and agreeable to almost all audiences. In contrasting three forms of objectivity—the views from nowhere, somewhere, and everywhere—I have shown how cultural understandings of objectivity in public reasoning are infused with different, often institution-specific, norms of deliberative rationality.[6]

A question that follows from all this is how the ideal of objectivity is securely maintained in the always contested, always deconstructed arenas of politics, where the contingency of the objective is continually on display. My second major contribution to the philosophies of science and politics addresses this puzzle, through the idiom of *co-production*.[7] Along with many others in science and technology studies, I believe that significant representations of the world, those new versions that we humans are al-

---

[5] On the situatedness of objectivity, see for example Lorraine Daston and Peter Galison, *Objectivity* (2007).

[6] For a long empirical elaboration of this point, see Jasanoff, *Designs on Nature* (2005).

[7] Key texts here include, from my work, Sheila Jasanoff, ed., *States of Knowledge: The Co-Production of Science and Social Order* (2004) and, for an empirical working through of the idea, *Designs on Nature* (2005).

ways constructing (see, for example, Nelson Goodman and Charles Taylor[8]), enfold normative as well as cognitive commitments. We believe, cognitively, that certain renderings of the world are the right ones because at the same time we believe, normatively, that it is right for the world to be as we have represented it. Each right makes the other right, even though the normative right may also need to be backed by considerable might in order to enforce the presumed correctness of its constructs. Much of my own academic work points to moments of emergence, contestation, and stabilization as periods of co-production, in which new versions of the world are constructed through the production of representations, institutions, discourses, and identities.

Political cultures, as my work shows, differ in the ways they accord credibility to stories told by public authorities, whether at an institutional or organizational level, at the level of nation states, or globally. What works persuasively in a court of law may not work so well in a regulatory expert committee; what counts as right reason in the United States may not be seen as sufficient to win acceptance in Britain or Germany. Evidence that the World Trade Organization or World Bank find compelling may not be seen that way in India, China, or Brazil. In puzzling through these disparities in the public uptake of reason, I have suggested that we need to take account of tacit cultural assumptions about constituting authority in public decisions. I call these clusters of assumptions *civic epistemologies*. These are interpretive rules pertaining to the production and evaluation of evidence, the assessment of public demonstrations or experiments, the meaning of objectivity, and the attribution of expertise in technical areas relevant to public actions. Civic epistemologies unite in a shared culture of reasoning those who act for the state and those who are affected by state action. These cultural practices reduce or eliminate the appearance of arbitrariness, even when actions or decisions are undertaken under conditions of epistemic uncertainty and lack of consensus. The rules of practice are sometimes tacit, as when they underwrite particular cultural understandings of objectivity and expertise, but they are sometimes explicitly articulated in law or longstanding administrative practice. As STS scholars take on board the all-important, often neglected reception side of the dissemination of scientific ideas in society, civic epistemologies will, I hope, become a topic of deeper philosophical interest and

---

[8] Taylor, 2007.

analysis.

4. What is the relation between philosophy of science and scientific practice, science policy, or efforts for social justice? Can there be a more productive relation? Is this desirable?

The short answer, of course, is yes: there should be closer relations, and such relations are desirable because they could be extremely productive. In general, I believe that philosophy of science could make important contributions to human understanding and human welfare by grappling with the epistemological issues that are embedded in, but not made explicit by, other areas of philosophical or social analysis that deal more centrally with questions of social justice.

For example, a task of political philosophy has been to uncover and make explicit the norms that govern just behavior, and thereby also to create the preconditions for increasing the total levels of justice in world society. We needed explicit philosophical work to make tractable the very notion of an "unjust war," even though traditions of warfare from ancient India to modern global conflicts have operated with basic understandings of what constitutes legitimate provocation for armed conflict, how prisoners of war should be treated, what protections exist for non-combatants, and which weapons are too nasty to use morally (Walzer, 1977). Underlying all such normative commitments, however, are epistemological foundations that are rarely explored by STS scholars, including those philosophers who have played a seminal role in strengthening the philosophical foundations of STS. Returning to an earlier point, for instance, it is clear that many of today's social justice issues are integrally bound up with questions about human nature and what is necessary for human flourishing. In turn, as I have suggested, those notions are undergoing reconsideration and redefinition as sciences and technologies reconstitute the meaning of the human. Such transformations and their implications should be among philosophy's core concerns.

Scientific practice presents another open door for renewed philosophical engagement with science studies and science policy more broadly. A generation ago, STS derived much of its force from arguing that the abstract concepts taken for granted by philosophers of science needed to be reexamined in the light of what scientists do rather than what they say. From the resulting turn to practice, there emerged an axiom of STS thought: truth, facticity, credibility, progress—all these are the endpoints of scientific work; only when the work of truth-making is finished do we have something that we can take to be the truth, or a believable fact, or an advance over earlier understandings. This insight has proved to be enormously liberating for ideas of social and political agency, since practice-based approaches show how people actually make the worlds they believe in. And yet principles and foundations have not entirely ceased to matter. There remains the social reality that people believe in principles, seek to articulate them precisely, and shape their arguments with reference to recognized norms. Foundationalist reasoning is itself a form of discourse and practice, and absent some dialogue with those practices, constructivist approaches focusing on science's material practices alone risk turning into mere behaviorist accounts. Elucidating what scientists do in the lab may reveal many pockets of unsuspected agency and indeterminacy, but such accounts do not explain why people use their agency in some ways rather than others. For this, we need more abstract and principled ways of thinking.

Globalization presents another domain for productive interaction between philosophy of science and broader normative questions. As I have indicated above, today's world presents many examples of rationalities in collision, even within and between closely similar societies. "Mad cow disease" strikes, and one man's meat becomes all too literally another man's poison. A global treaty decrees that countries should eliminate non-tariff barriers to trade, but it becomes clear that people are not always willing to take on board things, such as foodstuffs, that come imbued with alien cultural presumptions. Indigenous peoples contest the grant of intellectual property rights for technological products and processes extracted from knowledge that their communities once held in common. How should such epistemic conflicts be resolved in principled fashion? What will be the epistemic rights and entitlements of future global citizens? These are surely questions that ought to interest philosophers, and the results could prove highly consequential for the emerging cosmopolitan and transnational societies of the twenty-first century.

5. Where do you see the field of philosophy of science to be headed? What are the prospects for progress regarding the issues you take to be most important?

It may be more useful to talk about where I think the philosophy of science should be headed, since as yet I see little sign of movement in those directions. Perhaps the most general way to put it is that philosophy of science should become more daring about transgressing its own self-imposed intellectual boundaries. These include the analytic categories with which the field defines its sphere of action, as well as the divisions between philosophy of science and other related fields of study. Boundaries, after all, are made, not given, and there are rewards to be had from asking how they were made and what sustains them.

In the past, philosophy of science has made progress by narrowing its field of vision in much the same way that the natural sciences have advanced through reduction. Thus, philosophy of science has tracked discrete branches of science, such as physics or biology, separated science from technology as objects of philosophical attention, and stringently demarcated its objectives from those of other philosophical domains, such as ethics or political theory. Above all, philosophers of science have adopted as the external perimeters of their own questioning the same boundaries that scientists and engineers erect around their activities. Even the contemporary turn toward studying the material foundations of science, such as Hans-Jörg Rheinberger's "epistemic things," tracks mainly those things that laboratories make, not the more overtly hybrid kinds, from pre-embryos to fine particulate matter, that proliferate in the world through the complex interweaving of scientific, legal, and political thought and action (1997). Not surprisingly, philosophers of science have often felt more at home with scientists or with internalist historians of science than with social analysts of scientific activity.

My hope is that something of the complexity and dynamism that STS scholars have tried to capture through the idiom of coproduction will seep into the subject matter of philosophy of science. We need richer understandings of interactivity and change, through conversations between people who, with nose held high and eyes averted, cross the streets of intellectual discourse whenever possible to the theory side, and those others whose work seems to such purists to be too "applied" or too mired in the concerns of everyday life. STS scholarship has done much to problematize purity, and to muddy the taken-for-granted distinctions between fact and value, science and technology, metaphysics and epistemology, actor and object, evidence and argument, nature and culture. It is time for the philosophy of science to read that body of complicating work in earnest, and to admit that the lines philosophy finds so clear are frequently blurry in practice; and that clarity, when achieved, is not a matter of transcendental logic but is contingently derived from particular institutional histories and constellations of power.

This does not mean the end of philosophy of science—far from it. It just means that the teasing out of principle from the messiness of scientific and social practice will be every bit as difficult as extracting a line of pure poetry from the grossness of human experience. Yet it is altogether worth doing. As William Butler Yeats wrote in "The Circus Animal's Desertion," his moving reflection in old age on the deepest sources of poetic inspiration:

> Those masterful images because complete
> Grew in pure mind, but out of what began?
> A mound of refuse or the sweepings of a street,
> Old kettles, old bottles, and a broken can,...
> Now that my ladder's gone
> I must lie down where all the ladders start
> In the foul rag and bone shop of the heart.

References

Bloor, D. *Knowledge and Social Imagery*. London: Routledge and Kegan Paul, 1976.

Brickman, R., S. Jasanoff, and T. Ilgen. *Controlling Chemicals: The Politics of Regulation in Europe and the United States*. Ithaca, NY: Cornell University Press, 1985.

Callon, M. "Some Elements of a sociology of Translation: Domestication of the Scallops and the Fishermen of St. Brieuc Bay." In J. Law (ed.), *Power, Action, and Belief: A New Sociology of Knowledge?*, 196-233. London: Routledge and Kegan Paul, 1986.

Carson, R. *Silent Spring*. New York: Houghton Mifflin, 1962.

Daston, L. and P. Galison. *Objectivity*. Boston: Zone Books, 2007.

M. Gibbons, H. Nowotny, and P. Scott. *The New Production of Knowledge: The Dynamics of Science and Research in Contemporary Societies*. London: Sage Publications, 1994.

Gieryn, T. F. *Cultural Boundaries of Science: Credibility on the Line*. Chicago: University of Chicago Press, 1999.

Goodman, N. *Ways of Worldmaking*. Indianapolis: Hackett Publishing Company, 1978.

Gross, P. R., and N. Levitt. *Higher Superstition: The Academic Left and Its Quarrels with Science*. Baltimore: Johns Hopkins Press, 1994.

Hacking, I. *The Social Construction of What?* Cambridge, MA: Harvard University Press, 1999.

Nelkin, D. (ed.). *Controversy: Politics of Technical Decisions*, 3rd ed. Newbury Park, CA: Sage, 1992.

Nelkin, D., C. Hershey, and D. Mueller. "The Impact and Control of Technological Change." *Science Studies*. 4(1): 97-103, 1974.

Porter, T. M. *Trust in Numbers: The Pursuit of Objectivity in Science and Public Life*. Princeton, NJ: Princeton University Press, 1995.

Rheinberger, H.-J. *Epistemic Things*. Stanford: Stanford University Press, 1997.

Taylor, C. *Modern Social Imaginaries*. Durham, NC: Duke University Press, 2007.

Walzer, M. *Just and Unjust Wars*. New York: Basic Books, 1977.

Selected Bibliography

Authored Books

Designs on Nature: Science and Democracy in Europe and the United States. Princeton, NJ: Princeton University Press, 2005; Italian translation, Milan, Il Saggiatore, 2008.

Science at the Bar: Law, Science, and Technology in America, a Twentieth Century Fund book. Cambridge, MA: Harvard University Press, 1995; Italian translation, Milan, Giuffré, 2001.

The Fifth Branch: Science Advisers as Policymakers. Cambridge, MA: Harvard University Press, 1990.

Risk Management and Political Culture. New York: Russell Sage Foundation, 1986.

Controlling Chemicals: The Politics of Regulation in Europe and the U.S. (with R. Brickman and T. Ilgen), Ithaca, NY: Cornell University Press, 1985.

Selected Edited Books

Earthly Politics: Local and Global in Environmental Governance. Co-edited with M. L. Martello. Cambridge, MA: MIT Press, 2004.

States of Knowledge: The Co-Production of Science and Social Order. London: Routledge, 2004.

Handbook of Science and Technology Studies. Co-edited with G. Markle, J. Petersen, T. Pinch. Thousand Oaks, CA: Sage Publications, 1995.

Selected Articles

"Making Order: Law and Science in Action." In E. Hackett et al. (eds.), *Handbook of Science and Technology Studies*, 3rd ed., 761-786. Cambridge, MA: MIT Press, 2007.

"Biotechnology and Empire: The Global Power of Seeds and Science." *Osiris*. 21(1): 273-292, 2006.

"Technology as a Site and Object of Politics." In C. Tilly and R. Goodin (eds.), *Oxford Handbook of Contextual Political Analysis*, 745-763. Oxford: Oxford University Press, 2006.

"In the Democracies of DNA: Ontological Uncertainty and Political Order in Three States." *New Genetics and Society*. 24(2): 139-155, 2005.

"In a Constitutional Moment: Science and Social Order at the Millennium." In B. Joerges and H. Nowotny (eds.), *Social Studies of Science and Technology: Looking Back, Ahead*, Yearbook of the Sociology of the Sciences, 155-180. Dordrecht: Kluwer, 2003.

"Technologies of Humility: Citizen Participation in Governing Science." *Minerva.* 41: 223-244, 2003; reprinted in A. Bogner and H. Torgersen (eds.), *Wozu Experten? Ambivalenzen der Beziehung von Wissenschaft und Politik,* 370-389. Wiesbaden: Verlag fur Sozialwissenschaften, 2005; adapted and reprinted in C. Mitcham (ed.), *Encyclopedia of Science, Technology, and Ethics,* pp. xix-xxvi. New York: Macmillan Reference, 2005.

"Image and Imagination: The Formation of Global Environmental Consciousness." In P. Edwards and C. Miller (eds.), *Changing the Atmosphere: Expert Knowledge and Environmental Governance,* 309-337. Cambridge, MA: MIT Press, 2001.

"The Songlines of Risk." *Environmental Values.* 8: 135-152, 1999.

"The Eye of Everyman: Witnessing DNA in the Simpson Trial." *Social Studies of Science.* 28(5-6): 713-740, 1998.

"Science and Decisionmaking." (with B. Wynne and contributing authors), In S. Rayner and E. L. Malone (eds.), *Human Choice and Climate Change,* 1-87. Washington, DC: Battelle Press, 1998.

"Beyond Epistemology: Relativism and Engagement in the Politics of Science." *Social Studies of Science.* 26(2): pp. 393-418, 1996.

"Contested Boundaries in Policy-Relevant Science." *Social Studies of Science.* 17(2): 195-230, 1987.

11
Evelyn Fox Keller

Professor Emerita of the History and Philosophy of Science
Program in Science, Technology, and Society, MIT, USA

---

1. How were you initially drawn to philosophical issues regarding science?

I would say that an interest in philosophical questions is something of a primitive in my intellectual biography, and it was what first drew me to science—more specifically, to theoretical physics. That I did not remain in physics was more a consequence of vicissitudes of history and politics than of intellectual orientation. More specifically, I began my graduate work in theoretical physics at Harvard in the late 1950s—a time when women were almost entirely absent from, and, according to widespread beliefs at the time, "naturally" unsuited to this field. This may have been especially true at Harvard where any woman who aspired to be a theoretical physicist was looked upon as a curiosity, an anomaly, an aberration, and for whom any evidence of success was regarded with disbelief and indeed, denial (see Keller, 1977). In fact, it was an impossible situation, and I take my hat off to the very rare women whose commitment to the field managed to survive. Mine did not. Although I received my degree in theoretical physics, my dissertation was in molecular biology, a field that was then a far friendlier environment, and one particularly receptive to those coming from physics.

Molecular biology was certainly interesting, as well as challenging, but I did not find the experimental work it required well suited to my (seemingly constitutional) philosophical and theoretical leanings. I returned to theoretical physics for another try, eventually sliding into mathematical biology, but the next few years, even though undeniably productive (see, e.g., Keller, 1965; Keller and Segel, 1970), felt (both at the time and in retrospect) like a kind of intellectual wandering in the desert. Care for my two small children left academic work a part time pursuit, an occupation always available for the back burner. Indeed, it was the women's movement that opened up a path for me that would eventually re-ground my intellectual aspirations.

The year 1969-70 found me in California accompanying my husband on his sabbatical appointment, attempting to pursue my own work. But it was tough going. Science was turning into a plate of cold rice—not only for me, it seemed, but for many of my friends as well. Was this really a widespread a phenomenon? Curious about the answer, I set about collecting data on the careers of women in science, and the data clearly showed that it was indeed widespread. I wondered why this should be so.

But the question lay dormant for another few years. In 1974, I taught a course in women's studies (my first) and it provided an opportunity to reexamine my own career path from a political perspective. Doing so was revelatory. When I received an invitation to give a series of lectures on my work in mathematical biology at the University of Maryland, that same semester, I felt an imperative to (somehow) include my new insights into the lectures. And so I began my final lecture with a demographic (a birth and death) model of women in science that clearly displayed (on the basis of the data I'd collected in California) the precipitous attrition rates for women scientists after the Ph.D.. Devoting the remainder of the lecture to the question of why this should be so, I undertook a review of all the barriers that operate against women scientists. Perhaps the largest and most critical barrier, I concluded, was the widespread belief in a "natural" association between men and science, and an equally "natural" disassociation between women and science—that is, the view of science as intrinsically masculine. Where, I asked, does such a belief come from? What is it doing in science of all places? And what consequences has it had for the actual practice of science?

But these questions too lay dormant. It took a couple of years more for me to recognize them not only as legitimate questions, but as questions that I might be able to help answer. And so began my work on gender and science—an inquiry into the ways in which our understandings of science and objectivity have been shaped by their historical association with a particular ideology of gender (and more specifically, with a particular ideology of masculinity). Inevitably, such an inquiry led to a series of deeply philosophical and historical questions, taking me far a field from the kinds of scientific research I had been doing.

Perhaps, even, too far. I was, after all, a scientist, and scientists remained my primary audience. Thus, when physicists would say to me (as they often did), "You're not talking about science, you're talking about language—about how scientists talk, not about what they do," I took their complaint seriously. And I decided they were largely right. I had in fact already made some efforts to demonstrate the effects of pervasive gender metaphors on the science produced, but they were clearly inadequate. In fact, I realized that the force of language on science was a seriously understudied subject, and it definitely warranted sustained effort. And so my focus took yet another turn: from the study of gender and science to that of language and science. Indeed, this label might serve to characterize all the work I have done since.

2. What, in your view, are the most interesting, important, or pressing problems in contemporary philosophy of science?

Asked about "the most interesting, important, or pressing problems in contemporary philosophy of science," I can only respond personally—i.e., I can only try to say what "the most interesting, important, or pressing problems in contemporary philosophy of science" seem from my particular point of view.

For example, how *do* the ways in which scientists talk shape the science they produce? Questions about language and science have attracted increasing attention among philosophers of science in recent years, with a sizeable literature having now accumulated on areas of specific interest—e.g., on metaphor in science. But in my view, the subject needs a far more focused attention on the concrete ways in which language and action are intertwined. This requires, e.g., attention to ways (large and small) in which language shapes our conceptual frame, as well as to the specific ways in which words used in familiar domains carry expectations forged in those domains when they are employed (as they of necessity must be) in the charting of unfamiliar domains.

Indeed, language is a carrier of many kinds of expectations from one domain to another. Most obvious perhaps is the transfer of scientific/philosophical expectations from one field to another—as, e.g., in the ways in which expectations of atomic composition were carried from chemistry into many fields in the 19th century. But language can also serve as a carrier of social and cultural expectations—e.g., expectations about organizational structure, distribution of causal power, division of labor, bifurcation between passive and active elements, etc., etc.. An understanding of just how such expectations are carried over into our models of the natural world—based on detailed case studies—is, in my view, crucial for understanding the dynamics of scientific change, yet serious study of such questions remains in its infancy.

Shifting gears somewhat, and also slightly rephrasing the question, I suggest that many of the most interesting, important, or pressing problems in contemporary philosophy of science are coming from the life sciences. The reason for this is straightforward: Biological Science is currently in a state of extraordinarily rapid change and in some areas, of dramatic upheaval. Many of these changes require a rethinking of Biology's most basic concepts: What is a gene? Indeed, is the assumption that inheritance is particulate still appropriate? Relatedly, what is an organism? How does one get from a collection of molecules to the kind of functioning entity that has been the subject of natural selection? How do these molecules self-assemble and self-organize into a living cell? And finally, what is the difference between a live cat and a dead one? Many of these are old questions—if not settled, then long deemed unproductive. Yet today they are alive and well, in fact, at the forefront of a great deal of contemporary research. As such, they provide a wonderful opportunity for philosophers of science—both to help biologists clarify the questions at issue, and to rethink many of their own assumptions (many of which were formed in the life sciences of old, if not in the physical sciences)—assumptions, e.g., about causal dynamics, emergence, complexity, organization and, of course, function.

3. How has your work offered original contributions to discussion on science? What does your work reveal that others fail to appreciate?

I think my efforts to track the ubiquity of metaphors of gender in science, and to show that these metaphors, when taken in conjunction with associated ideologies of gender, had an important role to play in the delineation of "the new science," constituted—especially at the time I did this work—a fairly original contribution. And it is my hope that this work contributed to a sensitivity to these issues among philosophers and historians. I also think (or at least hope) that my focus on language more generally has helped to make philosophers and historians more conscious of the ways in which the richness and multifaceted nature of language helps shape the practice and conceptual frameworks of scientific practice. What I have found to be of particular importance in the construction of scientific arguments is the (surprisingly common) reliance on terms with multiple meanings (polysemy), where, as I have tried to show, chronic slippage between and among these various meanings proves to be crucial in sustaining arguments in a variety of contexts. Early on, I demonstrated the important role that slippage between ordinary and technical meanings of *competition* have played in mathematical ecology; more recently, I have attempted to track the conceptual support for simplistic (and outmoded) notions of genetic causation in molecular genetics that is provided by routine slippage between different referents of the term *gene*; and my current work on *The Mirage of a Space Between Nature and Nurture* argues that the persistence of this mirage owes a great deal to slippage between (a) genes and gene differences, trait and trait differences, and (b) slippage between technical and ordinary meanings of *heritability*.

The idiosyncratic nature of my career trajectory has also, I believe, made me sensitive to the particularities of what I call *epistemological cultures*—i.e., the collectivity of shared assumptions about what counts as an explanation, and about what suffices for understanding. Certainly, my unusually varied disciplinary experience has contributed greatly to my ability to identify and track such differences between disciplines, most especially, between the mathematical and life sciences.

4. What is the relation between philosophy of science and scientific practice, science policy, or efforts for social justice? Can there be a more productive relation? Is this desirable?

I will focus on the first part of this question, for it seems to me to be the most problematic. One might expect that important and mutually informative relations would exist between philosophy of science and scientific practice, yet my own experience suggests that it is not so. In my own work, I have benefited—even relied upon—extensive discussion with practicing scientists, but even though many scientists show interest in my work, frequently inviting me to speak at their conferences, it is my strong impression that my ideas and arguments only rarely filter down to their thinking about their own scientific practices. Furthermore, most scientists seem to show little interest at all—taking it for granted that nothing I have to say has any bearing on their work. Indeed, the suggestion that it might is often greeted with outright hostility. On more than one occasions when my work has been formally reviewed by working scientists, the review has concluded by warning the reader that "once again, we see that philosophers of science have nothing to teach scientists."

Obviously, I believe they are wrong, or I would not persist in my efforts to reach scientific audiences. Also, there is at least one area in which close interchange between philosophers of science and practicing scientists has had conspicuous benefits for both (I refer here to the field of population genetics where Dick Lewontin, R. has—almost single handedly—established a tradition of close collaboration). Furthermore, I myself have seen some instances (however rare they may have been) in which my work has been claimed to be useful to working scientists—in their formulation of their arguments, the posing of their questions, even the design of their experiments. Thus I cannot help asking, Why should so many scientists be so defensive? Why the insistence that they have nothing to learn from the philosophy of science? What elevates such a suspicion to a matter of policy?

Of course, I may be wrong—it may be that scientists have little if anything to learn from philosophers of science, but I suspect the question has not been put to a fair test. Indeed, I suspect that a good part of what is at work here is little more than good old territorial protectiveness. Despite my training, and earlier career, I am not now a working scientist. And my claim to the right (and the authority) to raise questions, challenge assumptions, engage in criticism—all in relation to a domain of activities of which I am no longer a member—seems, to these scientists, simply unacceptable. It threatens not so much the boundaries of their territory as their absolute authority over that territory.

I am not suggesting that scientists are unique in such behavior—far from it. I have also been challenged by historians for straying over disciplinary boundaries. But I regret such attitudes deeply, wherever they crop up. Perhaps this is why I never quite know what to call myself—am I a historian of science? A feminist scholar? A philosopher of science? A theoretical biologist? The answer seems to be yes, I am all of the above. I seem, again by constitution, to be a boundary crosser, and proud of it.

You ask, can there be a more productive relation? I think there can be, but only if there is extensive dialogue and engagement. But this is what I work toward, and continue to hope for.

Selected Bibliography

Books

A Feeling for the Organism: The Life and Work of Barbara McClintock. W. H. Freeman,1983; Second Edition, 1993.

Reflections on Gender and Science. New Haven: Yale University Press, 1985; Tenth Anniversary Edition, 1995.

Body/Politics: Women and the Discourses of Science. Co-edited with M. Jacobus and S. Shuttleworth. Routledge Press, 1990.

Conflicts in Feminism. Co-edited with M. Hirsch. Routledge Press, 1990.

Conversazioni con Evelyn Fox Keller, Elisabetta Donini, Eleuthera, Milan, Italy, 1991.

Keywords in Evolutionary Discourse. Co-edited with E. Lloyd. Harvard University Press, 1992.

Secrets of Life, Secrets of Death. Routledge Press, 1992.

Refiguring Life: Metaphors of Twentieth Century Biology, (Wellek Lectures). Columbia University Press, 1995.

Feminism and Science. Co-edited with H. Longino. Oxford University Press, 1996.

The Century of the Gene. Harvard University Press, 2000.

Making Sense of Life: Explaining Biological Development with Models, Metaphors, and Machines. Harvard University Press, 2002.

The Mirage of a Space Between Nature and Nurture. Duke University Press, in preparation.

Selected Articles

"Statistics of the Thermal Radiation Field." *Physical Review.* 139, 1B, B202. 1965.

"Slime Mold Aggregation Viewed as an Instability." (with L. A. Segel), *Journal of Theoretical Biology.* 26: 399415, 1970.

"The Anomaly of a Woman in Physics." In S. Ruddick and P. Daniels (eds.), *Working it Out,* 77-91. New York: Pantheon Books, 1977.

"The Mind's Eye." (with C. Grontkowski), In S. Harding and M. Hintikka (eds.), *Discovering Reality: Feminist Perspectives in Epistemology, Methodology and Metaphysics*. Reidel, 1983.

"Physics and the Emergence of Molecular Biology." *Journal of the History of Biology.* 23(3): 389-409, Fall 1990.

"*Drosophila* Embryos as Transitional Objects: The Work of Donald Poulson and Christiane Nüsslein-Volhard." *Hist. Studies in the Physical and Biological Sciences.* 26(2): 313-346, 1996.

"Is There an Organism in this Text?" In P. Sloan (ed.), *Controlling Our Destinies*. U. of Notre Dame Press, pp. 273-290, 2000.

"Models of and Models for: Theory and Practice in Contemporary Biology." *PSA.* 67: S72-S86, 2000.

"Models, Simulation, and 'Computer Experiments.'" In H. Radder (ed.), *The Philosophy of Scientific Experimentation*, 198-215. University of Pittsburgh Press, 2002.

"Revisiting 'Scale-Free' Networks." *BioEssays.* 27(1): 1060-1068, October, 2005.

"The Disappearance of Function from 'Self-Organizing Systems.'" In F. C. Boogerd, F. J. Bruggeman, J. H. S. Hofmeyr, and H. V. Westerhoff. (eds.), *Systems Biology*. Elsevier, 2007.

12
Philip Kitcher

Columbia University, New York, USA

---

1. How were you initially drawn to philosophical issues regarding science?

I came to philosophy of science from study of the history of science. My interest in philosophical issues about science was sparked by the provocative discussion of the growth of scientific knowledge offered by Thomas Kuhn in *The Structure of Scientific Revolutions*. Like many of Kuhn's readers in the 1960s, I focused on what I took to be a critique of the objectivity of science, and it seemed to me to be a central task to offer a picture of the growth of knowledge that would both be as sensitive to historical detail as Kuhn's treatment plainly was, and also recapture notions of rationality and progress. Only gradually did I come to appreciate the many ways in which *Structure* offered insights into scientific practice.

Reading Kuhn led me to a study of many other great figures in twentieth century philosophy of science: Carl Hempel, Rudolf Carnap, Ernest Nagel, Nelson Goodman, and Karl Popper. W. V. Quine's landmark essay "Two Dogmas of Empiricism" was also extremely influential. After an undergraduate career in which I had done mathematics for the first two years, and the Cambridge History and Philosophy of Science Tripos in my third year, I decided that I wanted to pursue philosophy of science at the graduate level. At the time, my background in philosophy was quite limited, although I had studied early modern philosophy of science (with Gerd Buchdahl). I suspect that the trajectory through which I came into philosophy of science (from mathematics through the history of science) has left its mark on my way of approaching philosophical questions.

2. What, in your view, are the most interesting, important, or pressing problems in contemporary philosophy of science?

Like most philosophers of science, a large part of my career has been occupied with the issues that were placed on the agenda by the pioneering logical empiricists (Hempel, Carnap et al.). I continue to think that those issues are most urgent when they are pursued in connection with the questions about the growth of scientific knowledge raised by Kuhn (and Paul Feyerabend and Stephen Toulmin). Yet I no longer believe that they are the most important topics for the philosophy of science. When reflective people consider the contributions of the natural sciences, they begin with the *value* of various forms of research. The ethical, social, and political questions about scientific practice have been sadly—I would say, absurdly—neglected in philosophical discussions. Today, more and more work in general philosophy of science (and even more in general philosophy) is undertaken on problems (often on puzzles) that are of interest only to a handful of specialists. It is impossible to defend this professionalization of philosophy by pointing to the example of the sciences—for the division of labor in philosophy is neither cooperative nor productive: philosophers do not build on one another's results, nor do they establish technical conclusions on which points of wider import can be based. The philosophers of earlier decades, from whose insights the present agenda for philosophy of science descends, were attacking serious issues, even if they were not concentrating on the ethical, social, and political questions I regard as central. Most contemporary discussions of issues about explanation, theory structure, confirmation, the character of laws, are quite irrelevant to any significant questions about the sciences.

By contrast, much of the work that has been done in recent years on the special sciences is both pertinent and illuminating. The great pioneers in the philosophy of physics (Hans Reichenbach, prominent among them) have been followed by generations of sophisticated researchers, who have offered many insights into relativity theory, cosmology, quantum mechanics, and thermodynamics. More recently, both the philosophy of biology and the philosophy of psychology have flourished, again focusing on problems that matter to the practitioners in these fields. Philosophy of economics and philosophy of chemistry are showing signs of following the same profitable trajectory.

I believe that there are general questions about science that continue to deserve philosophical attention. Philosophers have spent far too little time and effort considering the conditions that make for the advancement of *collective* or *community-wide* knowledge (again, this is a theme broached in Kuhn's pioneering work). Because professional studies in history and sociology of science have so frequently collided with philosophical work, the important enterprise of philosophical exploration of science in historical context has withered. Above all, philosophers have ignored the value-theoretic issues about science. What kinds of scientific research are valuable? How should scientific research be practiced within a democracy? What are the ethical constraints on scientific inquiry? Kuhn's *Structure* was widely read in part because it began to raise and address questions of this sort.

3. How has your work offered original contributions to discussion on science? What does your work reveal that others fail to appreciate?

During the past 35 years, I have offered a number of controversial suggestions about particular sciences and about the practice of the sciences generally. I began with a discipline that I had studied closely as an undergraduate: mathematics. From my first encounter with the philosophy of mathematics, I was convinced that the foundationalist pictures taken to be the leading candidates for philosophical understanding were wildly implausible. Taken seriously, all of them suggested that *real* mathematical knowledge only became possible in the late nineteenth century (at the earliest), thus condemning Euclid, Archimedes, Newton, Euler, Gauss, Galois, and Weierstrass to mathematical ignorance. In *The Nature of Mathematical Knowledge*, I attacked mathematical apriorism. My central proposal, and (I continue to think) my most important insight about mathematics, however, was the thought that the epistemological order recapitulates the historical order. The allegedly "foundational" pieces are justified because of their relation to the mathematical practices that preceded them.

In the philosophy of biology, I have spent much effort defending what all serious biologists believe, namely that Darwin's theory of evolution by natural selection was well established about 140 years ago, and has grown in strength ever since. I have done so because I thought it important for philosophers to rebut the popular impression that Darwinism is "just a theory." Philosophy should explain to people, as clearly and as patiently as possible, what is wrong with "Creation Science" and with its successor (or current avatar?) "Intelligent Design Theory." I have devoted two books (*Abusing Science*; *Living With Darwin*) to trying to discharge this task. In writing the second book, I have offered a thesis about the relation between science (broadly understood) and religion. I differ both from those who think that Darwinism is easily compatible with religion, and those who see a simple opposition. In my opinion, Darwin is simply the most visible figure in a many-sided case against *literalist* or *supernaturalist* religion. It follows from my account that we need a more humane secular humanism, one that may be able to find common ground with those religious traditions that abandon literalism/supernaturalism, going "beyond belief."

In the philosophy of biology, I have also defended some more controversial positions, arguing that human sociobiology, in its existing forms, is seriously flawed (see *Vaulting Ambition*). Although I have not given a full-dress version of the critique, I have extended the argument to consider contemporary Evolutionary psychology, which, although in some respects superior to the older sociobiology of the 1970s, recapitulates many of the same errors. To the surprise of some of my readers, I have been more sympathetic to claims about genic selection than most of my fellow-philosophers, and have suggested that decisions about the so-called units of selection are pragmatic and conventional. Here, I take seriously the thought that Darwin's metaphor of natural selection is just that: a metaphor. I have also defended pluralism about species concepts, and have opposed the most straightforward types of biological reductionism.

At the center of my work in the philosophy of science are two books that cover related themes. *The Advancement of Science* was my first attempt to wrestle with the issues I, like many philosophers, found so challenging in Kuhn's work. I hoped to develop an account of scientific progress and scientific rationality that would attend to the historical details, and yet also rebut the skeptical worries that many have read Kuhn as posing. In the course of doing this I tried to take some first steps towards considering the ways in which scientific communities, and not merely individuals, can be subject to epistemic appraisal. Here, I think, one can use the mathematical tools deployed in population biology and in economic theory to propose formal models of community-wide inquiry. It should be possible to articulate a set of methods for collective investigation that are as exact as some of those proposed for individual research, and to use those methods to evaluate various kinds of scientific institutions. One obvious question (although possibly not the most tractable) is to ask how much diversity of opinion is valuable in science, and to consider possible institutions that would support an appropriate amount of diversity.

Almost a decade later, I returned to some of the same general questions about science with a very different perspective. In the intermediate period, I had undertaken a study of the ethical, legal and social implications of the Human Genome Project, and had written a book (*The Lives to Come*) on these topics. That investigation convinced me that any adequate account of scientific progress and rationality had to be intertwined with a consideration of the value-theoretical questions that arise with respect to the science. In *Science, Truth, and Democracy*, I revised the general vision of science offered in *The Advancement of Science*, taking into account some of the criticisms presented by fellow-philosophers. Influenced by Nancy Cartwright, John Dupré, Ian Hacking, Martin Rudwick, and Peter Galison, I retracted my earlier commitment to philosophizing about a homogenous entity called "Science," and recognized the diversity of the sciences. This led me to see that the kinds of inquiries we pursue are always selective. Inspired by my experience with the Genome Project, I concluded that value-theoretic questions arise centrally for scientific research. To use a metaphor I find suggestive, scientific research is like map-making: it must begin from particular purposes, that set the content of what is to be represented, but, once those choices have been made, how the map is best drawn is a value-free matter. Our evolving interests are expressed in the scientific agendas we set—but the convenient first-person pronoun hides the important issue. Who should set this agenda? In the second part of *Science, Truth, and Democracy*, I argued that we needed an ideal of well-ordered science, a science in which the inquiries pursued are those that would respond to human needs in a fair and comprehensive way. Although I could not have phrased it in this way at the time, the book expressed my growing conviction that the central question in the philosophy of science concerns how to formulate standards for sciences that are to respond to human needs, broadly construed.

In recent years, my reading of the classical American pragmatists, Dewey in particular, has led me to a rethinking of the goals of philosophy. Dewey says, in a passage that (I imagine) would strike most Anglophone philosophers as utterly crazy, that "philosophy may even be defined as *the general theory of education.*"[1] He is, I believe, absolutely right. The center of philosophy lies in the identification of the ways in which human lives, individually and collectively, can go well, and the scrutiny of human practices to recognize how genuine values can be made available to all people. In the context of the sciences, that means understanding the contributions that scientific inquiry can make to human wellbeing, and a critical investigation of the scientific practices and institutions we have, from the perspective of this understanding. I now think that, although the ideal of well-ordered science was a good first step, philosophers need to go further, and to consider the ways in which that ideal can be implemented. In some recent work ("What Kinds of Science Should be Done?," "Global Health and the Scientific Research Agenda," "Scientific Research: Who Should Govern?," and "Does 'Race' Have a Future?"), I have been trying to take some further steps towards applying the ideal of well-ordered science. I suspect that most philosophers of science would regard these attempts as quite removed from the main questions of philosophy of science. By the same token, from my perspective, the puzzles that occupy so much of philosophical attention today are extraordinarily remote from the enterprise that Dewey envisages.

To pursue this project further, I have been led into areas quite remote from those to which I was introduced at the beginning of my career. I doubt that social and ethical questions about the sciences can profitably be discussed without a general approach to issues about values. Out of my work in the philosophy of biology came an interest in the implications of evolutionary theory for our picture of ethics, and, as I have pursued that interest, I have been led to develop an account of ethics (including normative ethics as well as meta-ethics) that is broadly naturalistic: I emphasize not only the evolutionary roots of ethical practice, but the social evolution and the historical processes that have shaped the forms of ethical life we now have—my views now owe as much (if not more) to Dewey as to (than) Darwin. This account serves as the basis for my current efforts to understand the social embedding of the sciences. It affords me ways of integrating the scientific picture of the world with responses to the concerns that have traditionally been satisfied by the world's religions. The closing chapter of *Living With Darwin* is a first step towards this integration—which

---

[1] See Dewey 1916, p. 328.

I see as putting the humanity back in secular humanism.
Whether this counts any more as philosophy of science does not seem to me a matter over which anyone ought to lose sleep. The importance of working out a satisfying scientific picture of the world and our place in it should be uncontroversial. That project is central to my current research.

4. What is the relation between philosophy of science and scientific practice, science policy, or efforts for social justice? Can there be a more productive relation? Is this desirable?

Following Dewey, I take philosophy of science to be intimately connected with scientific practice, with scientific policy, and with efforts for social justice. As already noted, much of the best work in contemporary philosophy of science is focused on problems that arise in the special sciences—*genuine* problems about our understanding of measurement in quantum mechanics or about potential biological causes of social behavior. Far less common on the contemporary scene are efforts to provide theoretical perspectives on science policy, or to consider whether scientific research programs promote social justice. Although some feminist philosophers have been in the forefront of taking these issues seriously, their work is often dismissed as "marginal." On the contrary, it should be viewed as central.

The major challenge for philosophy of science today is to find ways of developing a more productive relation between theoretical reflections on science and issues in social philosophy (including questions of science policy). Meeting this challenge will not be easy, since it requires philosophers of science to read bodies of literature that are not part of the traditional canon of the discipline. Fortunately, many philosophers of science have shown themselves adept at studying material in the special sciences. So there is hope that some will tackle the really central questions.

5. Where do you see the field of philosophy of science to be headed? What are the prospects for progress regarding the issues you take to be most important?

Some good things are happening in contemporary philosophy of science. Philosophy of physics continues to thrive, as do philosophy of biology, philosophy of psychology and philosophy of economics. Clark Glymour and his associates have shown how one can refine the causal-statistical methods needed in many areas of science. Nancy Cartwright has offered theoretical conceptions that can be used in pursuing issues about the human uses of science. John Dupré has been exploring the ways in which genomics has been shaped by contingent features of the societies in which it has developed, and how in turn genomics influences our conceptions of human life and human well-being.

Yet we still have far too little discussion of social and value-theoretic questions about science, too little interrogation of social epistemology, and far too many well-read, intelligent and imaginative people devoting years of their lives to solving refined problems that make no real difference. If philosophers of science were required to explain periodically why the research they are pursuing makes a real and important difference, I believe that we would see a renaissance of the subject, and some real progress on the central issues.

References

Dewey, J. *Democracy and Education*. The MacMillan Company, 1916.

Kitcher, P. "What Kinds of Science Should Be Done?" In A. Lightman, D. Sarewitz, and C. Dresser (eds.), *Living With the Genie*, 201-224. Washington D.C.: Island Press, 2003.

Kitcher, P. "Global Health and the Scientific Research Agenda." (with James Flory) *Philosophy and Public Affairs*. 32: 36-65, 2004.

Kitcher, P. "Scientific Research: Who Should Govern?" *Nanoethics*. 1(3): 177-184, 2007.

Kuhn, T. *The Structure of Scientific Revolutions*. Chicago: University of Chicago Press, 1962.

Quine, W. V. O. "Two Dogmas of Empiricism." *Philosophical Review*. 60(1): 20-43, 1951.

Selected Bibliography

Books

Abusing Science: The Case Against Creationism. MIT Press, 1982.

Vaulting Ambition: Sociobiology and the Quest for Human Nature. MIT Press, 1985.

The Advancement of Science. Oxford University Press, 1993.

The Lives to Come: The Genetic Revolution and Human Possibilities. Simon and Schuster (U.S.), Penguin (U.K.), 1996.

Science, Truth, and Democracy. Oxford University Press, 2001.

In Mendel's Mirror: Philosophical Reflections on Biology. Oxford University Press, 2003.

Finding an Ending: Reflections on Wagner's Ring. (with Richard Schacht), Oxford University Press, 2004.

Living with Darwin: Evolution, Design, and the Future of Faith. Oxford University Press, 2007.

Joyce's Kaleidoscope: An Invitation to Finnegans Wake. Oxford University Press, 2007.

Selected Articles

"A Priori Knowledge." *Philosophical Review.* LXXIX: 3-23, 1980.

"Explanatory Unification." *Philosophy of Science.* 48: 507-31, 1981.

"1953 and All That. A Tale of Two Sciences." *Philosophical Review.* XCIII: 335-373, 1984.

"Species." *Philosophy of Science.* 51: 308-333, 1984.

"The Return of the Gene." (with Kim Sterelny), *Journal of Philosophy.* 85: 335-358, 1988.

"The Division of Cognitive Labor." *The Journal of Philosophy.* 87: 5-22, 1990.

"The Evolution of Human Altruism." *The Journal of Philosophy.* 90: 497-516, 1993.

"Parfit's Puzzle." *Noûs.* 34: 550-577, 2000.

"Real Realism: The Galilean Strategy." *Philosophical Review.* 110: 151-197, 2001.

"Does 'Race' Have a Future?" *Philosophy and Public Affairs.* 35: 293-317, 2007.

13
Helen Longino

C.I. Lewis Professor of Philosophy
Stanford University, USA

---

1. How were you initially drawn to philosophical issues regarding science?

My first philosophical interest was the problem of reference. I pursued an MA in logic and epistemology at University of Sussex to try to sort out my questions about the relationship between our language and the world we sought to speak about. I ended up in philosophy of science after taking several philosophy of science courses with Peter Achinstein at The Johns Hopkins University. I saw that science, as a paradigmatic site for the use of referential language, provided a good focus for my philosophical interests. In the course of preparing to specialize in philosophy of science for my Ph.D., I also became fascinated by the history of sciences. My specific philosophical questions changed as I studied the philosophy of science more intensively, and a year under the tutelage of biologist Ruth Doell early in my professional career gave me important insights into scientific practice. Finally, I began to appreciate the profound interdependence of modern science and modern industrial and post-industrial societies. Understanding science promised to make a contribution to social philosophy as well.

2. What, in your view, are the most interesting, important, or pressing problems in contemporary philosophy of science?

I see several categories of problem in contemporary philosophy of science. There are projects to understand features of and episodes in the history of sciences, projects to identify and analyze the foundational concepts, principles, and presuppositions of various of the special sciences, projects to understand the logical and epistemological structures of scientific inquiry and the concepts used in analyzing those structures, projects to understand the relations between scientific inquiry and its social and cultural context. Some of these projects are interdependent and some of them work at cross-purposes. For example, understanding episodes in the history of sciences and understanding relations between inquiry and its context are mutually illuminating, and both either make assumptions about the logical structure of inquiry and theories or advance ideas about them. Identifying foundational concepts, principles, and presuppositions may be in conflict with an antifoundationalist account of the logic and epistemology of science. Because of this interdependence, I am reluctant to identify any problem as most pressing, interesting, or important. However, I think philosophers of science have an obligation to help nonphilosophers understand how to think about relations between science and society. Especially when important social and political matters turn on the results of difficult or controversial science, philosophers ought to try to make the methodological issues involved in evaluating competing theories and hypotheses clear to the general public.

3. How has your work offered original contributions to discussion on science? What does your work reveal that others fail to appreciate?

I see my main contributions as three-fold: 1) developing a rubric for understanding the role of values in scientific inquiry that neither assumes value-freedom nor leaves inquiry subject to political whim; 2) advancing a social epistemology for science; 3) showing how feminist ideas and scholarship could contribute to a philosophical understanding of the sciences.

1) The distinction between constitutive and contextual values makes explicit a distinction implicit in most discussions of science. It was conventional to suppose that one distinguishing feature of scientific inquiry was its value-freedom or value-neutrality. Of course only some kinds of values were those from which science was free: social, political, or aesthetic values, which were thought to compromise the truth-seeking aspect of inquiry. On the other hand, truth itself is a value, and as the understanding of hypothesis evaluation became more complex, philosophers of science included other qualities, such as simplicity and generality, as relevant to evaluation. These were claimed to be internal to

science, or definitive of science. In *Science as Social Knowledge* I proposed that values thought to be internal to or definitive of science be labeled constitutive values. Other values that might play a role in scientific thought I proposed be called contextual values, to indicate that they are not internal to science, but belong to the context in which scientific inquiry is pursued. These values might be aesthetic, might be pragmatic, or might be socio-political. I made several points regarding this distinction. First, it is a functional, rather than a categorical distinction. Values are constitutive when they play a constitutive or internal role in inquiry, that is, when they are used in evaluation of hypotheses or research programs. Contrary to philosophers and others who thought that activity qualifying as scientific could be distinguished from that classifiable as non-scientific by its exclusion of contextual values, I argued that sometimes values originating in the context of inquiry played a constitutive role. That is, sometimes, contextual values limited or informed the background assumptions of particular research programs that were generally acknowledged to be scientific research programs. They either motivated the emphasis placed on other values, such as simplicity, or entered more directly in the evaluation and interpretation of data. Whether or not this was so had to be established on a case by case basis. I should emphasize that my claim was not a general claim, but a possibility claim, based on what I took to be the logical openness of evidential relations.

2) These ideas about values in science are incorporated into a more general epistemological approach to scientific knowledge. I was led to develop what I call Critical Contextual Empiricism by three lines of thought: 1) that theoretical hypotheses are underdetermined by the evidence available for them, 2) that scientific judgments often reflect extra-evidential considerations (via Kuhn, Feyerabend, and analysis of particular research programs), 3) that scientists nevertheless take empirical evidence to be some ultimate court of judgment as well as priding themselves on the openness of science.

I began thinking about the first of these points in connection with studying Carl Hempel's analysis of confirmation. In an article published in 1979 ("Evidence and Hypothesis," in *Philosophy of Science*), I argued that this concept could not be made to apply to the most interesting cases of scientific inference. Treating confirmation as the deductive derivability of data statements from the development of a hypothesis for a universe consisting only of the entities referred to in those data statements could apply to situations in which hypotheses were generalizations of data statements. This analysis could not, however, be applied in situations in which hypotheses contained terms that referred to entities and relationships that were not included in descriptions of the data. Some further hypotheses or assumptions were required to establish evidential relevance in these cases. In inquiry, we are seeking to understand the processes that produce the phenomena we can observe and measure. Establishing generalizations about observables is just establishing the evidence base for hypotheses, models, and theories that can be used in explanation. Thus, I argued, data alone were never sufficient to support one among a set of contesting hypotheses if only because their evidential relevance to any hypothesis was a function of background assumptions establishing (or affirming) the relationship between data and hypothesis.

This dependence on background assumptions bore a family resemblance to positions advanced by Paul Feyerabend, Russell Hanson, and Thomas Kuhn. They all relied on detailed case studies of episodes in the history of science to argue that multiple theories were compatible with available data at periods of theory choice. However, they also chose to account for this by reference to views of the theory-ladenness of meaning. I argued that theory-ladenness did not do justice even to the historical episodes they invoked in its support. In addition, it made of theories hermetically sealed systems invulnerable to criticism. Treating data statements, theoretical hypotheses, and background assumptions as independently meaningful made possible logical as well as critical analysis of evidential relationships. It also made the practices of scientists intelligible. Scientists succeeded in communicating disagreements (and agreements) across theoretical commitments, so meaning holism was an inapt way to account for the historical evidence of multiple theories explaining the same phenomena.

I later came across the work of Pierre Duhem, discovering to my pleasure and chagrin a view of evidential relations very akin to mine. But Duhem, faced with the problem of justifying background assumptions, could only appeal to the physicists' good sense ("bon sens"). I had gone in a different direction.

How could dependence on (undischarged) background assumptions be compatible with claims of objectivity for science? I argued that we needed to think of scientific inquiry and method as practiced by communities, not by individuals. Looked at this way, critical interaction within and between communities of researchers becomes an integral aspect of method. Descriptions of data, methods of measurement, conceptual soundness of hypotheses, reasonableness of background assumptions are all subject to critical examination. I proposed that a community's research practices be regarded as objective to the extent that its critical interactions satisfied conditions of venue, uptake, publicity of standards, and (tempered) equality of intellectual authority. The outcome of a community's process of choosing among competing hypotheses could be regarded as objective to the extent that the critical scrutiny of competitors satisfied those conditions. This was a process that could eliminate idiosyncratic values and assumptions, but it made evident how so-called contextual values could function as constitutive ones. If they were shared by all members of the community and played a role in data classification and description or in hypothesis elimination or selection, they functioned as traditionally understood constitutive values.

Understanding data classification and description as independent of the theories for which they served as evidence made it possible to understand how scientists could disagree about the proper interpretation of observations (they assigned them relevance in light of different background assumptions). It also makes it possible to recognize the priority scientists ascribe to observation and measurement as arbiters of theory. The description and classification of data is not independent of all theory, but is not wholly determined by the theory for which it serves as evidence. While not immune to criticism, observational and experimental data serve as the least defeasible element in a theoretical argument. The empiricism of Critical Contextual Empiricism consists in this priority given to observational and experimental data. Including criticism as an aspect of method also has support from scientific practice. However imperfect, peer review is an essential aspect of scientific practice, as is the emphasis on reproducibility of experiments. Peer review is the contemporary descendent of the exchange of ideas that characterized the development of modern Western science in the 17th century. I see this critical interchange referenced to some set of commonly agreed standards of evaluation as being as important to the development of modern science as the development of experimental methods.

In my second book, *The Fate of Knowledge*, I deepened this social analysis. Instead of arguing for the addition of social interaction to extant understandings of scientific method, I argued for the social, interactive character of the cognitive practices of science themselves, for the social character of observation and of justificatory reasoning. In addition I introduced a concept, *conformation*, to replace *truth* as the term of evaluation of theories and hypotheses. True/false, I followed other philosophers in arguing, is too crude an evaluative rubric and does not apply to a significant amount of scientific content. In some cases what is desired is similarity, approximation, isomorphism, homomorphism, or some other semantic relation. conformation is the catch-all term intended to cover all forms of semantic success. Truth is one, but not the only, form of such success. Because degree and respect of conformation must be specified in order to fully spell out the standard of success, social interaction will be implicated here also. This notion of conformation also introduces an element of pragmatism, as the degree and respect of conformation required will depend on the purposes for which the knowledge is sought.

The three main advantages of Critical Contextual Empiricism are that it avoids problems associated with earlier empiricist accounts of theory evaluation and acceptance while preserving the empiricism of scientists, incorporates the lessons of Kuhn and Feyerabend about multiplicity without falling into holism, and avoids foreclosing the metaphysical issue of monism/pluralism. This social account is not relativist, but because it holds that empirical data constrain without necessarily limiting to a single theory, it is open to scientific knowledge having, in the end, either a monist or pluralist character.

3) Finally, I believe this socializing of scientific knowledge has made evident the general relevance of feminist work in the philosophy of science. I used the analysis of values and of evidential relations to show how androcentric perspectives did and could persist in particular scientific research programs, notably physical anthropology and behavioral neuroendocrinology. The role of these perspectives did not disqualify the research as science, but was a legitimate subject of examination and critique, especially when that was accompanied by alternative accounts of the available evidence. Additionally, I showed how some feminist critiques of scientific research programs could be understood as critiques of the prima facie non-contextual values shaping inferences in those programs. Both substantive and methodological choices, then, can be affected by androcentric (and, by the same token, gynecentric) values, and feminist analysis contributes to the critical evaluation of both substantive and methodological aspects of science.

4. What is the relation between philosophy of science and scientific practice, science policy, or efforts for social justice? Can there be a more productive relation? Is this desirable?

Some philosophers of science think our task is to explicate norms of an ideal science. I think this results in a distorted conception of science, the promulgation of unrealizable ideals that can be used to discredit perfectly good science, and, ultimately, disenchantment. [I think the extreme social constructivist views that were popular briefly are an example of such disenchantment.] I think our normative task is a bit more complicated: to articulate epistemically or cognitively normative notions in a way that can be applied to science as practiced and that take into account the conditions under which humans seek knowledge. This requires a back and forth relation between conceptual analysis and empirically informed accounts of scientific practice. Philosophy of science doesn't produce methodologies to be employed by scientists, rather it helps us understand the consequences of employing particular methodologies.

Science policy is often produced under the guidance of a presupposed philosophy of science. If science policy has as a goal of organizing the funding of science so that political decisions requiring information have the best quality information available, then philosophical ideas about the nature of knowledge and inquiry will be an inevitable background to policy. Philosophy of science ought to be able to make those ideas and their presuppositions explicit and, when appropriate, criticize them. For example, monism about scientific knowledge or absolutism about evidential relations might lead a policy maker to think it doesn't matter who does science; good method will lead to the best results. A contextualist may think that researchers with different experiences or metaphysical/ontological presuppositions may come up with equally empirically acceptable accounts of a process. When equally good alternatives are in contention it is important to understand the limitations and prospects of each of them and this in turn requires promoting diverse approaches, and by implication a diverse research community.

I do not think there is any intrinsic relationship between philosophy of science and social justice. Depending on one's view of scientific knowledge and methodology, social justice concerns may or may not enter into science, or may enter in one phase, but not in another. I don't thereby dismiss the question, but think its answer depends on one's prior philosophical views. And what constitutes social justice, how we defend claims about social justice, are themselves controversial questions in social and political philosophy. I do, however, think it is important to think about the relations between science and social justice. This leads me to the last of the 5 questions.

5. Where do you see the field of philosophy of science to be headed? What are the prospects for progress regarding the issues you take to be most important?

Philosophy of science continues to make progress in the various areas mentioned in my answer to question 2. The best of philosophy of science will be concerned with understanding developments at the forefront of the special sciences, with bringing analysis of normative and methodological concepts into alignment with their functions in scientific inquiry, and with refining the analysis of relations between science and society. Since this is an area of particular interest to me, I am very pleased to note the increasing attention it is receiving from philosophers. Although I disagree with certain central claims in Philip Kitcher's *Science, Truth, and Democracy*, I can only applaud its publication and his opening the discussion for mainstream philosophers of science of the problems of governance of science in a democracy. Miriam Solomon is examining the epistemology of consensus conferences in medicine. Nancy Cartwright is examining the logic of evidential arguments in educational policy. Heather Douglas has written about values in toxicology research. Elisabeth Lloyd is examining the structure of evidential relations in climate change research. James R. Brown is investigating issues in pharmaceutical research. Several edited volumes and sessions at Philosophy of Science Association and American Philosophical Association meetings have been devoted to relations between science and society and between philosophy of science and society. These efforts extend the work of philosophers concerned to analyze the structures of research concerned with (alleged) race, sex, and class differences. It is by no means a comprehensive list. But it shows the range of issues to which philosophers of science can usefully apply ourselves. And it augurs well for increasing attention by philosophers to these important questions.

References

Ankeny, R., and L. Parker. (eds.) *Mutating Concepts, Evolving Disciplines: Genetics, Medicine, and Society.* Kluwer Publishing, 2002.

Duhem, P. *The Aim and Structure of Physical Theory.* Princeton University Press, 1954.

Feyerabend, P. "Explanation, Reduction, and Empiricism." In Feigl and Maxwell (eds.), *Minnesota Studies in the Philosophy of Science, vol. 3.* University of Minnesota Press, 1962.

Hanson, N. R. *Patterns of Discovery.* Cambridge: Cambridge University Press, 1958.

Hempel, C. "Studies in the Logic of Confirmation." In C. Hempel, *Aspects of Explanation.* The Free Press, 1965.

Howard, D., J. Kourany, and M. Carrier. (eds.) *The Challenge of the Social and the Pressure of Practice.* University of Pittsburgh Press, 2008.

Kitcher, P. *Science, Truth, and Democracy.* Oxford University Press, 2000.

Kuhn, T. *The Structure of Scientific Revolutions.* Chicago: University of Chicago Press, 1962.

Longino, H. "Evidence and Hypothesis." *Philosophy of Science.* 46: 35-56, 1979.

Longino, H. *Science as Social Knowledge.* Princeton: Princeton University Press, 1990.

Longino, H. *The Fate of Knowledge.* Princeton: Princeton University Press, 2002.

Machamer, P., and G. Wouters. (eds.) *Values, Objectivity, and Science.* University of Pittsburgh Press, 2005.

Selected Bibliography

Books

Science as Social Knowledge: Values and Objectivity in Scientific Inquiry. Princeton University Press, 1990.

Readings in Feminism and Science. Co-edited with E. F. Keller. Oxford University Press, 1996.

Women, Gender, and Science, (*Osiris*, vol. 12). Co-edited with S. Gregory Kohlstedt. University of Chicago Press, 1997.

The Fate of Knowledge. Princeton University Press, 2002.

Scientific Pluralism, a volume in the Minnesota Studies in Philosophy of Science series, Co-edited with S. Kellert, and C.K. Waters. University of Minnesota Press, 2006.

Selected Articles

"Values, Heuristics, and the Politics of Knowledge." In D. Howard, J. Kourany, and M. Carrier (eds.), *The Challenge of the Social and the Pressure of Practice*. University of Pittsburgh Press, 2008.

"Circles of Reason: Feminist Reflections on Reason and Rationality." *Episteme*. 2:1, Feb. 2006.

"Evidence in the Sciences of Behavior." In P. Achinstein (ed.), *Evidence*. Johns Hopkins University Press, 2005.

"Behavior as Affliction: Framing Assumptions in Behavior Genetics." In R. Ankeny and L. Parker (eds.), *Mutating Concepts, Evolving Disciplines: Genetics, Medicine, and Society*. Boston: Kluwer Publishing, 2002.

"Social Dimensions of Scientific Knowledge" for *Stanford Encyclopedia of Philosophy*, (invited and refereed entry posted April 2002, revised August 2006) http://plato.stanford.edu/entries/scientific-knowledge-social/

"What Do We Measure When We Measure Aggression?" *Studies in History and Philosophy of Science*. 32(4): 685-704, 2001.

"Explanation v. Interpretation in the Critique of Science." *Science in Context*. 10(1): 113-128, 1997.

14
David Papineau

Department of Philosophy, King's College London, UK

---

1. How were you initially drawn to philosophical issues regarding science?

My first degree in the mid-1960s was a BSc at the University of Natal (now KwaZulu-Natal). I did a range of courses, including some in physics and economics, but my main focus was on mathematics and mathematical statistics. I can remember feeling philosophical niggles about some of the topics I was introduced to, especially the conventional account of statistical inference, but that wasn't why I decided to switch to philosophy. It was just that I acquired a range of broader interests. I wanted to know how things worked generally, especially how people and society worked. It was a close call between philosophy and psychology, but I'd started reading Bertrand Russell and A.J. Ayer, and they made me excited about philosophy.

However, when I went to Cambridge to read philosophy for a second undergraduate degree, I naturally gravitated back towards topics in philosophy of science and mathematics. At that time Cambridge required philosophy undergraduates to specialize in their final year. Either you did all logical and technical topics, or you did all ethical and historical topics, but you couldn't mix them up. It was an odd system, and the authorities must have thought so too, for they changed it soon afterwards. But for better or worse my final year was devoted exclusively to philosophy of science, philosophy of mathematics, mathematical logic, and philosophical logic. So when it came to choosing a PhD topic, it was natural enough for me to stay in this general area and revert to my earlier interest in statistical inference.

Of course this was very much a Cambridge topic, with R.A. Fisher and Richard Braithwaite the founding fathers and Ian Hacking and Hugh Mellor their younger disciples. I signed up to work with Hacking, who had recently published his first book *The Logic of Statistical Inference* (1965). But in the end I did my thesis on something quite different, though still with Hacking. This was the era of Kuhn and Feyerabend, of incommensurability and 'anything goes,' and I was intrigued and excited. In the 1960s we were going to make everything new, and Kuhn and Feyerabend showed how this was possible within science. Science could help us to discard the old ways of seeing and open up new possibilities.

It was the radicalism of Kuhn and Feyerabend that attracted me, not the relativism many took them to be advocating. I certainly didn't buy the line that all theories are equally good. Indeed my doctoral thesis was devoted to showing how evidence-based theory-choice could survive even radical meaning incommensurability.

What I didn't realize, however, was that the resulting position (published as *Theory and Meaning* 1980), was only marginally different from relativism. At that time Popperstill dominated philosophy of science, at least in Britain, and we took it for granted that all scientific theories were fated to be falsified. Of course this simply amounts to a sceptical denial that science gets at the truth. But we weren't supposed to think this, and it took me an embarrassingly long time to see through the emperor's new clothes. (I was greatly helped by attending some gloriously irreverent lectures by David Stove in Sydney, later published as *Popperand After*, 1981.)

I spent a great deal of the 1980s figuring out how to be a real realist about science. Many different positions attract the label 'anti-realism' and for each of them there is an opposed position worth calling 'realism.' I am in favour of pretty much all these realisms, but they raise different issues and require different arguments.

I worked through all this in my *Reality and Representation* (1987). Since then I have spent less time on the epistemology of science and more on its metaphysics. In a way this has been a return to my original motivation for taking up philosophy. I've become less interested in examining the credentials of scientific theories, and more interested in what their better-confirmed parts tell us about the world we live in.

2. What, in your view, are the most interesting, important, or pressing problems in contemporary philosophy of science?

Well, for a start there are the problems I myself work on. Of course

I don't think they are important because I work on them—but I do like to think I work on them because they are important. At the same time, I'm not so megalomaniac as to suppose that I work on *all* the important problems. (I'd like to, but sadly life is too short.) So let me first talk about the problems I do work on, and then I'll say a little bit about some other issues.

Mind and Matter
Most of my work over the past couple of decades has been concerned to assess the manifest image of everyday thought against the scientific image. In particular, I have asked how much of our ordinary conception of the human mind can survive within the scientific picture. (See especially my *Philosophical Naturalism*, 1993.) Consciousness and representation are the two tricky issues here. If there isn't anything more to us than the physical nature uncovered by the natural sciences, then how are feelings and mental representation so much as possible?

If these questions aren't interesting and important, then I don't know what is. And I would say that the general outlines of the right answers are pretty widely accepted, at least in the philosophical circles I move in. Conscious states are nothing over and above physical states, though our intuitive ways of thinking about consciousness can conspire to conceal this from us. And representation is a general biological phenomenon, with human thought merely the most sophisticated species of the genus of cognitive representation that is widespread in the biological realm.

Still, even if these general answers are agreed, I would say that there remains plenty of work to be done. We may have a general idea of how to locate minds within the scientific image, but it shouldn't be taken for granted that all of the manifest image of the mind will be preserved when we do so. Science may still show that many of our more specific ideas about consciousness and human intentionality are misplaced. Indeed neuroscience already suggests that conscious states play a much less central role in cognition than we normally suppose, and that intentional states are much less sentence-like than is assumed by common-sense psychology. The important issues facing the philosophy of psychology over the next few decades will be at this level: as science finds out more and more about our cognitive mechanisms, how many of our detailed everyday assumptions about the mind will survive?

Special Sciences
Some of my recent work has addressed a more general aspect of the relation between manifest and scientific images. It is fairly

widely supposed that there can be 'special sciences' in the sense of disciplines that are not type-reducible to physics but nevertheless display lawlike patterns of their own. I have always been doubtful about this. Note that it is in-principle reducibility that is at issue here. We can all agree that there are law-rich sciences that are not *practically* reducible to physics. We need look no further than chemistry for a clear example. But advocates of 'special sciences' want to make the stronger claim that there can be law-rich sciences which aren't even type-reducible *in principle*, because their kinds are 'multiply realizable' in a way that chemical kinds clearly aren't. This seems quite implausible to me. In my view kinds can only enter into a serious system of laws if they have a common physical basis. (Thus, for example, we might have a serious set of laws about primate vision, given the physical commonality between primate eyes, but not about animal vision in general.)

If this is right, then an urgent task is to figure out which kinds do have a common physical basis and which do not, the better to know what we can expect of them. With kinds of the former sort, we can hope to find sets of systematic laws. Kinds of the latter sort may still be usefully invoked in identifying causes and explaining particular facts, but they won't give rise to systematic law-rich sciences. It seems to me an open question, for example, how much of our ordinary thinking about people and human societies fits into which category. Sometimes we will be referring to psychological categories with a common physical basis, and thus the potential to underpin laws, while at other times we will be referring to mental and social kinds that are indeed variably realized at the physical level, and so unsuitable for serious laws. Those who are interested in the possibility of an economic science, say, will do well to consider how far they are working with categories of the former rather than the latter kind. (Cf Papineau, forthcoming a.)

Causation
Philosophical worries about the manifest and scientific images can involve the scientific image as well as the manifest one. What exactly does science tell us about the world we live in? There is plenty of philosophical work to be done in making sense of scientific ideas.

A central issue here is the nature of causation. This plays a crucial role in many sciences, yet seems to escape the grasp of fundamental physical theory. Basic physics deals in temporally symmetric processes which are unsuited to account for the temporally directed relation of causation.

Much recent philosophical work has followed David Lewis in seeking an explanation of causation in terms of counterfactuals. This strikes me as multiply misguided. For a start, we need causation to explain counterfactuals, not the other way round, which renders Lewis's programme circular. Moreover, even if we take counterfactuals as given, cases of preemption block any equation of causal with counterfactual relations. And finally, it is questionable whether counterfactuals have the directionality needed to ground causal asymmetry.

When Lewis addressed this last problem, he appealed a 'de facto' actual-world asymmetry, the asymmetry of overdetermination. However, if we are going to appeal to some further such actual-world symmetry, why not use it to analyze causation directly, without the detour through counterfactuals? It is not clear that Lewis's own asymmetry will do the trick, but there are others on offer. I myself have long been interested in the possibility of reducing causation to the kind of asymmetries displayed by correlational facts. (See Papineau, 2001b.) For example, when two uncorrelated event types are both correlated with a third, we can be confident that the third is an effect of the first two, rather than vice versa. This is just the kind of fact assumed by 'Bayesian net' methods for inferring casual structures from correlations. Few experts on Bayesian nets are prepared to commit themselves to a reduction of causation to correlational structures, but in fact their techniques open the way to just such a reduction. Dan Hausman's book *Causal Asymmetries* (1998) does a very nice job of showing exactly what assumptions we need to reduce causes to correlations.

Even if we accept the general outlines of such a reduction, much still remains unclear. One issue is to understand the kind of correlational facts that might provide a reductive basis. In the first instance, correlations are identified using observation and statistical inference. But these correlations also need some kind of modal status if they are to be serious candidates for reducing causation. It seems to me an open question exactly what kind of modal status is required here. Related to this are questions about the relation between type and token causation. Correlation-based causal facts are not only general, but have a kind of generality that abstracts away from the full details of basic physical processes, akin to the generality displayed by thermodynamic laws. This makes me think that the relation between causal structure and token physical processes is rather different from what we intuitively suppose.

Quantum Mechanics and Everett

Another central puzzle about the scientific image is the interpretation of quantum mechanics. I first became interested in this when I was in Cambridge in the 1980s, and Michael Redhead, Jeremy Butterfield and Mary Hesse were at the centre of debates on the topic. By the end of the decade I was a convinced Everettian. The 'many-worlds' approach pioneered by Everett is certainly weird, but all the alternatives struck me as far worse. Non-local causation, preferred reference frames, conspiratorial initial conditions,... I didn't see how anybody with any feeling for the physical truth could take these options seriously.

At one level the Everettian line couldn't be simpler. The world is governed by Schrödinger's equation alone, without 'wave collapses' or 'vector state reductions.' Decoherence then explains how reality is constantly evolving into effectively independent macroscopic branches. But of course the resulting picture is so radically different from orthodox thinking that huge areas of our world-view will need to be reworked if Everett is right. Philosophers interested in Everettianism have made a start, focusing in particular on the role of probability in rational choice, but much remains to be done. What exactly is the foundational status of objective probability, if it is a measure over the 'approximate worlds' created by decoherence? What are we referring to when we talk of tables, chairs, or people, if such ordinary persisting objects are constantly branching? Are regret and relief appropriate reactions to unlikely outcomes, if we know that on most branches of reality these outcomes did not occur? And so on.

Temporal Asymmetries

Moving away now from my own areas of interest, the other general field that seems to me most in need of attention from philosophers of science is the direction of time. Perhaps 'the direction of time' is a misnomer here—what is really at issue are processes that display an asymmetry *in* time. I have already talked about causation as one such process. But it is by no means the only one. There is also the familiar thermodynamic asymmetry of entropy increase, and the not quite so familiar asymmetry of retarded rather than advanced radiation. Moreover, quantum mechanical decoherence and the associated evolution (that is, renormalization) of probabilities are themselves temporally orientated processes. And if we switch from the cosmic to the anthropocentric, we find the asymmetry of knowledge and the asymmetry of action—we know about the past in a way we don't know about the future, and we can influence the future in a way we can't influence the past. All these phenomena are asymmetric in the sense that they conform to generalizations that make essential reference to a preferred orientation in time from 'earlier' to 'later.'

It seems likely that at some level these temporally orientated phenomena all have a common basis. Certainly some have argued that the asymmetry of entropy and the asymmetry of radiation both stem from the initial low entropy state of the universe. Still, none of these temporally asymmetric phenomena is as yet entirely understood even on its own, let alone in relation to the others. It would be a great advance if we had a coherent account of the nature and interrelations of all these different temporal asymmetries.

Clarity on these matters is likely to pay off in other areas as well. Much recent work in the foundations of physics has focused on various kinds of apparently non-local and temporally reversed dependencies. At first sight these may seem anomalous. But I suspect that much of the appearance of anomaly arises only because different levels of analysis are being conflated. Constraints involving locality and temporal order are appropriate to temporally directed phenomena, but not necessarily to processes at more fundamental physical levels. Once we are clearer about the nature of temporally directed processes, we will know better when such constraints must be respected and when not.

3. How has your work offered original contributions to discussion on science? What does your work reveal that others fail to appreciate?

Much of my past work has straddled the border between philosophy of science and more mainstream topics in metaphysics and philosophy of mind. To keeps things simple, let me restrict my answers to this question to work that falls squarely within the philosophy of science.

Realism

As I said earlier, I have defended realism about scientific theories against various different 'anti-realisms.' This has involved a number of different components. In particular, I have defended a theory of representation that allows that even our best theories can in principle fail to be true—in opposition to such anti-realist semantic doctrines as verificationism, semantic instrumentalism, and Peircean pragmatism. And I have defended an epistemology that implies that, even so, our theories are beliefworthy—here in opposition to the quite different species of epistemological anti-realism that set the standards for knowledge unreasonably high.

Many philosophers of science are impatient with these foundational niceties, but I think that clarity on these matters is essential to a proper understanding of the arguments for and against 'scientific realism.' We cannot possibly deal accurately with the 'no miracles argument,' the 'pessimistic meta-induction,' the 'underdetermination of theory by evidence,' and so on, unless we are clear about the nature of representation and knowledge. I work through these issues in a lot of my writings, but I think that the best illustration of the importance of foundational issues is the Introduction to my edited *Philosophy of Science* (1996), where care about the basics allows me to unravel a range of confusions in the realism debate and show where the real issues lie.

Theoretical Terms

Over the last decade, I have been revisiting the very first topic I worked on—'the theory-dependence of scientific terms.' Back in the 1970s, we had the following problem. Given that meaning is use, and that theoretical commitments affect the use of scientific terms, and that there's no difference between the analytic and synthetic parts of theories—given all that, it followed that *any* change in a theory altered the meanings of all its terms. We struggled with this argument and wondered where it left scientific objectivity. But nothing was ever really decided. Instead Putnam's and Kripke's causal theories of reference came along to tell us that the meanings of scientific terms have nothing to do with surrounding theories, and we moved onto other issues with a sigh of relief.

However, this now strikes me as having been too quick. Even if the Putnam-Kripke line works for some scientific terms, there are good reasons to think that other terms really are theory-dependent. Apart from anything else, surely we want to allow that the fate of at least some scientific terms is bound up with that of the surrounding theory—'phlogiston,' 'ether,'. . . Moreover, as these examples suggest, such theory-dependence remains important to issues of scientific realism. Fortunately, we are now in a position to make progress.

Looking back, I am nowadays struck by how much of the old 1970s debate was contorted by an unthinking verificationism ('meaning is use'). If we drop the verificationism, then we can have theory-dependence without meaning variance or incommensurability. Moreover, the troublesome analytic-synthetic arguments cease to carry any cosmic consequences about the indeterminacy of translation, reference, and everything else, and instead simply point to a benign species of vagueness in certain scientific terms. Philosophers of science who still think that theory-dependence is somehow hugely problematic—and I fear they are not uncommon—may not realize how far they are still thinking in verificationist terms. Anyway, I have worked through these issues in a number of papers, especially 'Theory-Dependent Terms' (1996), and I hope that these make it clear how clear-headed non-verificationists can stop worrying and learn to live with theory-dependence.

Physicalism

I've written quite a lot about the rationale for metaphysical physicalism—that is, for the thesis that all facts reduce to (or at least supervene on) the physical facts. Philosophers of science aren't always keen on this kind of reductive thesis. They are apt to point out that most physical theories don't even reduce their historical predecessors, let alone anything else. But of course this historical observation isn't to the metaphysical point, important as it may be for conditioning our epistemological attitude to current physical theories. The metaphysically interesting issue isn't about relationships between *transient theories*, but between *eternal levels of reality*. Looking at it from a God's-eye point of view, is there a basic level that fixes what happens at other levels?

Around the 1950s, philosophical fashion started moving strongly in the direction of a positive physicalist answer to this question, under the influence of figures like Feigl, Smart, Putnam and Davidson. I found the position attractive, but the sudden burst of enthusiasm seemed to count as much against physicalism as for it. If physicalism was such a good idea, why wasn't there a longer tradition of advocacy? The lack of any historical precedents suggested that the excitement might owe more to mindless twentieth-century science-worship rather than serious argument.

This got me looking at the arguments actually on offer, and it quickly became apparent that the crucial assumption was the 'causal completeness of physics.' All the philosophers mentioned above took it as given that any *physical* effect must have a full *physical* cause. Physicalism quickly follows, via the thought that mental, or biological, or any other putatively non-physical facts then won't be able to influence the physical world unless they are themselves included among those physical causes.

But in a way this only pushed the problem back. If the completeness of physics was such a good idea, then in turn why hadn't everybody always believed *that*? And the answer to this turned out to be very interesting. As I explained in "The Rise of Physicalism" (2001a) and again in the Appendix of my *Thinking about Consciousness* (2002), the completeness of physics isn't some kind of conceptual principle, but a highly empirical hypothesis on which informed scientific opinion has changed its mind a number of times. Most importantly, while the mechanical philosophy of the seventeenth century upheld the completeness of physics, orthodox Newtonian mechanics was unequivocally committed to denying it, via its enthusiasm for vital and mental force fields. And this commitment happily survived the nineteenth-century discovery of the conservation of energy, given the then standard assumption that these non-physical force fields would themselves be conservative. It was only well into the twentieth century that detailed physiological research persuaded scientific thought against vital and mental forces, thereby making the completeness of the physical realm available for philosophical arguments in favour of physicalism.

Everett and Probability

As I explained above, Everettianism seems by far the most attractive interpretation of quantum mechanics on offer. It requires no sudden collapses, non-local action, or experimental conspiracies. Physical systems simply conform to Schrödinger's law. But many philosophers and physicists feel that Everettianism undermines itself. In particular, they are worried that Everettianism seems unable to account for probability and its relation to rational decision. Of course, Everettians have quantities that behave numerically like probabilities, namely, the squared moduli of the amplitudes of the branches' wave functions. But the sceptics complain that these Everettian quantities lack features of probability—in particular relations to uncertainty and frequencies—that are essential to the way that probability guides action.

When I first heard this objection, from Euan Squires in the early 1990s, I scarcely took it seriously. I felt like Yul Brynner in The Magnificent Seven when the undertaker tells him the dead Indian can't be buried because there's no driver for the hearse—"Oh, hell! If that's all that's holding things up, I'll drive the rig." It struck me as easy to show that Everett has no special problems with probability.

I still feel this, but now realize that many others don't see things the same way. I attribute the difference to my having worried a lot about probability before I ever came across Everett. I used to bang my head against the rationale for the Principal Principle (in effect—what's so good about betting with the objective odds?) and I still have a draft of half a book on rationality in a drawer that I never finished because I became so tangled up on this issue. I eventually came to the view that there just isn't any way to justify the Principal Principle—surprisingly, it is a primitive feature of rationality that you ought to bet with the objective odds, even if this isn't guaranteed to get you what you want.

So when I was told that Everettianism can't explain why probability should guide rational action, I felt this was a case of the pot calling the kettle black. Orthodox metaphysicians can't explain this either. Indeed, as I argue in a number of papers, Everettianism looks positively better off on this issue (Papineau, 2004; forthcoming b). At least Everettians are guaranteed to have chosen the best action if they bet with the odds. Once Everettians maximize objective expected utility over all future branches, they are no longer hostage to the possibility that they may nevertheless have made the wrong choice—as rational agents they still are on the orthodox assumption of one actual future. (It was this that most persuaded me of Everttianism in the first place—it's always seemed horrible to me that orthodoxy should primitively require us to perform actions that may nevertheless turn out to have been the wrong choice.)

Some Everettians respond to the probability challenge by trying to construct an Everettian justification for the Principal Principal (Deutsch, 1999; Wallace, 2003). This strikes me as the wrong move, especially given that proposed justifications seem to assume most of what they promise to prove. Everettians should simply say that orthodoxy can't give any justification either, and moreover that this is far more embarrassing for orthodoxy that it is for Everettianism.

4. What is the relation between philosophy of science and scientific practice, science policy, or efforts for social justice? Can there be a more productive relation? Is this desirable?

I'm not sure that philosophy of science does have much advice to offer scientific practitioners or policy makers. It is true that such people will often need to know how far the evidence supports some specific theory (say, that carbon dioxide levels are causing global warming) or indicates that some experiment is worth performing (will the Large Hadron Collider detect the Higgs boson?). Still, there is no reason to suppose that philosophers of science are particularly well-placed to answer such practical questions. To a large extent the issues they address are of a far more general nature, and even when they do engage with specific scientific subjects their opinions need carry no more weight than those of other experts. I don't want to say that scientific researchers are the only people entitled to form a view when their subjects raise important practical issues. Sometimes these researchers will be too close to the issues, and a healthy society will contain a wider range of people who are competent to assess the evidence. Of course, this wider body may well include philosophers of science—but if so it won't be because of their special philosophical skills, but simply because they are among those who know something of the relevant subject.

5. Where do you see the field of philosophy of science to be headed? What are the prospects for progress regarding the issues you take to be most important?

As I have said, I am nowadays rather more interested in the metaphysics than the epistemology of science. Perhaps this puts me in a minority among contemporary philosophers of science. Certainly there has been an explosion of work over the past couple of decades on the nature of scientific practice, and in particular on the models and other kinds of intellectual constructions that scientists use to represent reality. At one level I am all in favour of this. It is certainly progress to realize how far actual scientific theorizing is from the traditional picture of deductively organized systems of law statements.

However, I do sometimes worry about the widespread tendency to read metaphysical conclusions straight off from these methodological advances. Just because practicing scientists treat causes as prior to correlations, or take the direction of time as primitive, there is no reason for metaphysicians to endorse these attitudes. Unthinking scientific lore is no sure guide to metaphysical truth. There are many reasons why scientists might find it easier to think of the world in ways that are unfaithful to its metaphysical structure.

I also worry in this connection about the increasing divorce between philosophers of science and mainstream metaphysicians. This unfortunate division is long-standing in Great Britain, where the classically-trained Oxford philosophers have traditionally been happy to leave science to the Popperians. But through most of the twentieth century the United States were different, with figures like Carnap, Hempel, Quine and Putnam as central to mainstream philosophy as anybody. More recently, though, the American philosophy of science community has been growing and defining itself in opposition to mainstream philosophy. This trend helps nobody and I can only hope it is soon reversed. The philosophers of science need the theories of the mainstream metaphysicians to help them make sense of the world that science uncovers, and the mainstream metaphysicians in turn need the philosophers of science to tell them about the actual nature of things.

References

Deutsch, D. "Quantum Theory of Probability and Decisions." *Proceedings of the Royal Society of London.* A455: 3129-37, 1999.

Hacking, I. *The Logic of Statistical Inference.* Cambridge University Press, 1965.

Hausman, D. *Causal Asymmetries.* Cambridge University Press, 1998.

Stove, D. *Popper and After.* Pergammon, 1981.

Wallace, D. "Everettian Rationality: Defending Deutsch's Approach to Probability in the Everett Interpretation." *Studies in the History and Philosophy of Modern Physics.* 34(3): 415-439, 2003.

Selected Bibliography

Authored Books

For Science in the Social Sciences. Macmillan, 1978.

Theory and Meaning. Oxford University Press, 1980.

Reality and Representation. Blackwell, 1987.

Philosophical Naturalism. Blackwell, 1993.

Introducing Consciousness. Icon Books, 2000

Thinking about Consciousness. Oxford University Press, 2002.

The Roots of Reason: Philosophical Essays on Rationality, Evolution and Probability. Oxford University Press, 2003.

14. David Papineau

Edited Books

The Philosophy of Science. Oxford University Press, 1996.

Essays on Teleosemantics. Co-edited with Graham Macdonald, Oxford University Press, 2006.

Selected Articles

"Introduction." to my edited *Philosophy of Science.* Oxford University Press, 1996.

"Theory-Dependent Terms." *Philosophy of Science.* 63: 1-20, 1996.

"The Rise of Physicalism." In C. Gillett, and B. Loewer (eds.), *Physicalism and its Discontents.* Cambridge University Press, 2001a.

"Metaphysics over Methodology—Or, Why Infidelity Provides No Grounds to Divorce Causes from Probabilities." In M.-C. Galavotti, P. Suppes, and D. Costantini (eds.), *Stochastic Causality.* Chicago: University of Chicago Press, 2001b.

"David Lewis and Schrödinger's Cat." *Australasian Journal of Philosophy.* 82(1): 153-169, 2004.

"Can any Sciences be Special?" In C. Macdonald, and G. Macdonald (eds.), *Emergence.* Oxford University Press, forthcoming a.

"A Fair Deal for Everettians." In J. Barrett, A. Kent, S. Saunders, and D. Wallace (eds.), *Many Quantum Worlds?* Oxford University Press, forthcoming b.

15
Stathis Psillos

Professor
Dept of Philosophy and History of Science, University of Athens, Greece

1. How were you initially drawn to philosophical issues regarding science?

It was a mixture of science and politics that drew me into philosophy of science during my years as a physics undergraduate. To me, and I guess to other leftists and ex-marxists of my generation, the early 1980s were a period of crisis of belief. A relatively neat conceptual scheme by means of which the world was viewed quite solid in its basic principles despite the many substantive differences in detail from thinker to thinker and from country to country—was collapsing. The hope that the world can change for the better via mass political action was coming to grief—at least for me. Contrary to Karl Marx's famous eleventh thesis on Feuerbach, the point was still to interpret the world—before we try to change it. The world cannot change unless *we* change; unless we take as basic and unnegotiable some principles of rationality and the search for truth and justification. Science was, to me, the bastion of rationality and progress; the *terra firma* upon which one could base all hopes for a better world. I believed back then, and still believe now, that science is the best way we have invented to push back the frontiers of ignorance and error, to achieve a deep understanding of the world and of our place in it, and to make the world a better place to live. But science and its claim to truth and knowledge needed justification and defence. This was a demand we inherited from the philosophers of the enlightenment—to subject to criticism even the most dearest and seemingly unassailable of our beliefs. This demand became even more topical in the age of modernity—where science itself was being dramatically transformed, delivering immense benefits to humanity, but also inflicting misfortunes. To me, looking into the scientific realism debate was no longer optional. I came into this debate with no neutrality. I wanted to defend scientific realism, along with the objectivity and rationality of science and its method. This was both an intellectual and a political goal. Back in the 1990s, there was a pervasive thought, especially among left-wing American intellectuals, that undermining the alleged epistemic authority of science, challenging its claims to objectivity and knowledge, was an act of liberation from the strangling authority of Reason—perhaps from capitalism itself. I was never persuaded by this rhetoric. It conflated intellectual authority with authoritarianism and, at least to all of us who learned our basic politics in the European south, intellectual authority (and objectivity and criticism and the search for truth) were the arch enemies of any kind of authoritarianism.

2. What, in your view, are the most interesting, important, or pressing problems in contemporary philosophy of science?

Philosophy of science has changed a lot during the twentieth century. If I were to offer a thumbnail summary of its course during the last century I would put it as follows. It started in the noughties with huge crises in the sciences and mathematics and its agenda was shaped by philosophically minded scientists who were battling for the prospects of certain theoretical ways to view the world and for competing views about the limits and scope of science and its theories. It took the form of the logic of science in Vienna in the early 1930s, where formal methods were employed to analyse and explicate the basic concepts of science and its method. 'Metaphysics' became a dirty word, but objectivity (mostly in the guise of structural invariance, or the common-factor point of view) was still what was sought after. It took the naturalist turn in the USA of the 1950s, leaving behind 'first philosophy' and its own allegedly special method of conceptual analysis and a priori insight in favour of a view that philosophy is continuous with the sciences. It looked for history-of-science-compliant macro-models of science and its growth in the 1960s, but it soon became apparent that little useful to the several individual sciences could be said at this level of generality and abstraction. The 1970s saw an explosion of interest in the philosophy of the individual sciences (in particular in the so-called special sciences—biology, psychology, economics etc.). The metaphysics of science staged an impressive comeback in the 1980s, with its full panoply of issues: causation, laws of nature, necessity, properties, natural kinds and the like.

But what has been most impressive is that in the last quarter of the twentieth century there has been an enormous diversity in the agenda of philosophy of science—formal methods (led by an increasing interest in Bayesianism); social studies of science; cognitive models of science; computational philosophy of science; feminism and gender studies; ethical issues in science and others. The result of all this is that it is hard nowadays to share interests with more than a few other people that work in cognate areas and even harder to stay in touch with philosophy of science as a whole.

Overall, the movement in the philosophy of science in the twentieth century has been from a more globalist approach to science and its problems, exemplified in a concern with broad philosophical issues about science (such as the aim and structure of theories, the nature and limits of explanation, the relation between the 'rational' and the 'factual', as Ernst Cassirer put it) to a more localist conception of philosophical problems, where philosophy of science is seen primarily as a toolbox for fixing problems that arise in the sciences. I cannot help feeling there is a certain loss in this movement of thought—we nowadays pay so much attention to the trees that there is too little to spare for the wood. There may well be different conceptions of what the wood is, but in my own opinion, the wood is still the globalist agenda that animated philosophy of science in the beginning of the twentieth century. This is partly because I have a Sellarsian account of what philosophy should aim at. For me, philosophy of science is an attempt to start with a determinate conception of reality as described by our best science and to try to understand *it*, what the world is like according to it, how it came about, and what it implies for us and for the ways we know and transform the world.

I would single out the following issues as most important or interesting.

- The role of mathematics in scientific theories; in particular, the fact that it seems (*prima facie* at least) that the abstract and the concrete are interwoven in our scientific conception of the world. Mathematics seems epistemically indispensable for understanding the world, but it is metaphysically suspect (or so many think) because mathematical entities are causally inert. Reworking the relation between the abstract and the concrete in physical theory will perhaps open up new ways to conceptualise the world.

- Do scientific theories imply any substantive commitments about the deep metaphysical structure of the world? This is an area where the most exciting work that has been going on in the philosophy of the individual sciences (physics, chemistry, biology, psychology, economics) can come in creative contact with the most exciting research that has been taking place in analytical metaphysics of science. Analytical metaphysics of science is empty without the philosophies of the individual sciences; the philosophies of the individual sciences are blind without analytical metaphysics of science.

- The relation between the epistemic and the ethical. Science is subject to epistemic norms but its increasing relevance to the lives, well-being and prosperity of people (and of other animals, and of the planet as a whole) requires that it is subjected to ethical norms too. There is need for a new deal between the epistemic and the ethical in science and hence a sustained development of an ethics of science.

- Competing conceptions of objectivity and rationality. Here too, there is space for a creative interaction between the formal work that has been done on logical and probabilistic models of rationality and decision-making and the strongly emerging view that values play an ineliminable (yet not algorithmically determined) role in scientific judgement.

- Philosophy of science should keep looking into its past aiming to better understand the thought and theories of the thinkers and schools of the twentieth century and before.

3. How has your work offered original contributions to discussion on science? What does your work reveal that others fail to appreciate?

It is hard to talk about your own work and its originality. This is best judged by the others (and, ultimately, by posterity). However, false modesty is no less a vice than arrogance! My own work has really fallen into two stages (and a break in between). I thought hard about scientific realism for about a decade and produced my *Scientific Realism: How Science Tracks Truth* (1999). Then, I had a break from realism, working mostly on causation and explanation. In the last few years, I have come back to scientific realism, aiming to rework and rethink the way its defence was articulated and advanced in my 1999 book. This reworking appears in my Knowing the Structure of Nature: Essays on Realism and Explanation (2009) and in some forthcoming papers, which re-examine the relationship between realism and empiricism. Part of the motivation to rework my commitment to realism has come from my venture into metaphysics. I have the highest respect for metaphysics, but I want to have as little of it as possible (it's an illusion, I think, to believe that you can leave metaphysics behind altogether).

I think two are the most distinctive marks of my work on scientific realism. The *first* has to do with the abductive defence of realism. Ever since Hilary Putnam put forward the slogan that realism 'is the only philosophy of science that does not make the success of science a miracle,' the defence of scientific realism has been based on what has come to be known as the 'no-miracles' argument. This argument has had quite a long history and a variety of formulations—some of which can be traced as early as in the beginning of the 20th century. In my view, the structure and role of the no-miracles argument in the realism debate is quite complex and my own research has aimed to unravel this.

The way I read it, the no-miracles argument aims to defend the reliability of scientific methodology in producing approximately true theories and hypotheses. Following more concrete types of explanatory reasoning that occur all the time in science, it suggests that it is reasonable to accept certain theories as approximately true, at least in the respects relevant to their theory-led predictions. These successful instances of explanatory reasoning in science provide the *basis* for a grand abductive argument. The no-miracles argument, however, is not just a generalisation over the scientists' abductive inferences. Although itself an instance of the method that scientists employ, it aims at a much broader target: to defend the thesis that Inference to the Best Explanation is reliable. Nowadays, however, I have somewhat changed my mind. I tend to think that the no-miracles argument is an argument within the realist framework and not an argument for it.

One central objection to the no-miracles argument is that it is *viciously* circular. But the abductive defence of realism proceeds within a broad naturalistic framework, within which the charge of circularity loses most of its bite because it is not justification of inferential methods and practices (at least in the neo-Cartesian internalist sense) that is sought but their explanation and defence (in the epistemological externalist sense). What I added to this defence, based on a well-known (though controversial) distinction between premise-circularity and rule-circularity (a premise-circular argument employs its conclusion as one of its premises; a rule-circular argument conforms to the *rule* which is vindicated in its conclusion), is that (a) the abductive defence of realism is rule-circular, (b) rule-circularity is *not* vicious, and (c) rule-circularity is involved in the defence of all basic rules of inference.

The *second* mark of my work on scientific realism has to do with my arguments against the pessimistic induction. The thought here is that the history of science is replete with theories that were once considered to be empirically successful and fruitful, but which turned out to be false and were abandoned. If the history of science is the wasteland of aborted 'best theoretical explanations' of the evidence, it might well be that current best explanatory theories will take the route to this wasteland in due course.

In order to reconcile the historical record with realism, I have claimed that realists should be more selective in what they are realists about. This led me into some work on particular past theories (like the caloric theory of heat and the nineteenth-century optical ether theories) aiming to show that those parts of them that essentially contributed to their empirical successes were retained in subsequent theories. This is what I have dubbed the *divide et impera* move. A claim that has emerged with considerable force is that theory-change is not as radical and discontinuous as the opponents of scientific realism have suggested. Realists ground their epistemic optimism on the fact that newer theories incorporate many theoretical constituents of their superseded predecessors, especially those constituents that have led to empirical successes. The substantive continuity in theory-change suggests that a rather stable network of theoretical principles and explanatory hypotheses has emerged, which has survived revolutionary changes, and has become part and parcel of our evolving scientific image of the world.

Critics of my views have raised a number of important objections. I have learned a lot from them—though so far I have not responded to most of them in an orderly and systematic way. There is, I think, a substantial issue of disagreement between me and many of my critics, which I take it to have broader implications for the way we view science and its relation to evidence and truth. I am an anti-holist in matters of confirmation and the *divide et impera* move requires anti-holism. Most of my critics rely on holism to challenge my arguments. Anti-holism can be easily misunderstood. Holistic theories of confirmation back in the 1960s did play a crucial role in the defence of scientific realism, since it was on their basis that it was shown that theoretical assertions (assertions that make claims about typically unobservable entities) are no less confirmable than observational ones. Evidence clearly goes 'all the way up' to the remotest theoretical reaches of the theory—it does not stay to what to the theory says about the observable entities. But in the heat of the battle, it was not sufficiently stressed that though theoretical assertions are confirmable, they are not all equally well-confirmed by the evidence; nor are all of them equally contributing to the successes of the theory; nor even to its explanatory potential.

I now think that when we (philosophers) think about scientific theories and what they assume about the world, we need to balance two kinds of evidence. The first (let's call it *first-order evidence*) is whatever detailed and specific evidence there is in favour (or against) a specific scientific theory—evidence that has to do with the degree of confirmation of the theory at hand. The second kind of evidence (let's call it *second-order evidence*) comes from the track-record of scientific theories and/or meta-theoretical (philosophical) considerations that have to do with the reliability of scientific methodology. This second-order evidence feeds claims such as those that motivate the Pessimistic Induction. In assessing scientific theories and science as a whole, these two kinds of evidence need to be balanced. How exactly this balance should be stricken is an interesting philosophical issue. It seems that it will be a contextual matter. Philosophers of science can help specify what kinds of factors and considerations can determine the context. What is also important is that there should not be double standards in confirmation, based on a supposedly principled distinction between OK-entities and not-OK ones. (This distinction is drawn along several lines, most typically between empirical and theoretical entities or between observable and unobservable entities or between entities to which there is independent epistemic access and entities to which there is not.) If no absolute privilege is conferred on either the first-order or the second-order evidence that is brought to bear on theories and if there are no double standards of confirmation of supposedly distinct parts of theories, a variety of epistemic stances towards scientific theories might be enunciated, depending on the context. All this might pave the way for a rapprochement between contextualist versions of instrumentalism and scientific realism.

4. What is the relation between philosophy of science and scientific practice, science policy, or efforts for social justice? Can there be a more productive relation? Is this desirable?

Philosophy of science is in relative isolation from scientific practice. This is both good and bad. It is good because philosophy of science is primarily *philosophy* of science. Hence, it is not a commentary on what scientists do; nor is it exclusively dependent on what trends and views there are in current science. There could not be philosophy of science without science—but the subject matter of philosophy of science is *philosophical* problems that arise within science. The current scientific worldview ought to act as a constraint on philosophy of science and on philosophical thinking in general. It will be hard to defend a view on strictly philosophical grounds, especially if it leads to consequences that are in conflict with what our best theories tell us about the world and our own place and role in it. This, however, does not imply that a philosophical conception of science, let alone a philosophical stance about what the world is like, is *dictated* by current science. Naturalism will be trivialised and will trivialise philosophy if it is pressed too far.

The bad thing is that philosophy of science, at least occasionally, is practised in such a way that there seems to be no caring at all about scientific practice; nor any concern about getting the scientific facts right. Abstract philosophical argument is good, but should make some contact with the real world. A pertinent example is some of the discussion about the argument from the underdetermination of theories by evidence. This concerns a deep philosophical problem and has been discussed thoroughly ever since Descartes formulated it in a relatively precise way with his evil-demon hypothesis. The spectre of scepticism has to be reckoned with and this calls for a thorough re-evaluation of what we take to be knowledge and justification (and of what resources, including empirical ones, we use to address these issues). There has been some exciting work in epistemology here in the last two decades. But I have the feeling that some bad philosophy of science has been fostered in relation to this issue: an exercise in philosophical imagination attempting to devise empirically equivalent alternatives to scientific theories (abstractly understood) which are totally uninteresting from a scientific point of view and are motivated by purely philosophical doubts and considerations. Taking account of actual science can help philosophers come down, every now and then at least, from the platonic heaven.

Science policy is an area in which philosophers of science ought to play a more central role. In science policy there is typically conflict between competing norms and priorities. A number of decisions are based on a mixture of facts and value-judgements. Philosophy of science can help create a balance (which, again, can be context-dependent) between facts, values and interests. Philosophical analysis (and inter-disciplinary input) can lead to the creation of practical frameworks: (context-dependent) norms of action in different cognitive areas. What is particularly important is that philosophy of science can offer insights and formal tools concerning decision-making under uncertainty and risk-management. It can explore ways in which epistemic responsibility and social responsibility reinforce each other. It can act as the mediator between what is sometimes seen (by the educated public) as the dogmatism of science and what is sometimes seen (by scientists) as social prejudice. Some other issues that philosophers of science can be particularly helpful (and in which their work might have a broader impact) are: managing scientific controversies and expert disagreement; examining the sources of uncertainty; finding new uses for evidence and for values in science policy making; managing social controversies (e.g., related to the public perception of scientific controversies, risk-perception etc.). The common factor in all this is that good science policy requires a better understanding of what science is and how it works.

I wish philosophy of science could have a stronger role to play in issues of social justice—or to promote social justice, for that matter. I think feminist philosophy of science and in particular my own favourite feminist standpoint epistemology has done much good in promoting social justice while securing claims to objectivity and truth. But I still believe that the cause of social justice is best served when we leave our papers and research projects every now and then and engage in local social action. Nowadays, the keyword seems to be 'excellence.' And excellence we do need—but we should not forget that we should also create opportunities for excellence where they are most needed.

5. Where do you see the field of philosophy of science to be headed? What are the prospects for progress regarding the issues you take to be most important?

Progress in philosophy is hard to measure. Perhaps, the emergence of consensus is a sign of progress—at least we come to agree that some problems are not worth exploring any more or some positions are no longer viable (though by no means dead). There seems to be not much going today for reductive understandings of the meaning of theoretical terms, for strong instrumentalist accounts of scientific theories, for purely formal models of explanation, and for global type-reductive accounts of inter-theoretic relations. Perhaps, the whole philosophical issue of conceptual change and the alleged incommensurability is pretty much exhausted.

Devising grand theories of science, of the form that dominated the 1960 and 1970s seems to have run out of steam. There are some rich ideas connected with these grand models (for instance, the Kuhnian idea of a normal science or the Lakatosian idea of a progressive research programme). But the deep and interesting work that has been going on in the foundations of the individual sciences has tended to highlight the disunity in science. There is room and need, I think, for a new synthesis. The diversity of the sciences can be looked at again, this time with an eye to common structures (especially in methods and in the patterns of explanation—e.g., in terms of mechanisms). This may well lead to cross-fertilisation—both in form and in content.

It's a good sign that the metaphysics of science has become a hot topic provided that this will be a scientifically informed metaphysics. But I am sceptical about the emergent neo-Aristotelianism. The barren landscape painted by the empiricist philosophers (based as it was on the thought that the regularities there are in the world do not need metaphysical enforcers) is being redrawn, this time with the full Aristotelian panoply—active powers, essences, necessary connections and the like. To my mind, these are unexplained explainers, and though everyone has to accept *some* unexplained explainers, in this particular case, they are more poorly understood than what they are supposed to explain. I favour a Wittgensteinian attitude here: "a nothing could serve just as well as a something about which nothing could be said" (*Philosophical Investigations*, §304).

After a period of what has been called 'the new fuzziness,' formal methods in the philosophy of science have come back. This time the horizons have considerably expanded. New formal tools are being used, from non-classical logics to probability theory and game theory. There has been a lot of exciting work on causal modelling and causal inference using Bayes networks; on ampliative and defeasible reasoning (with a lot of input from Artificial Intelligence); and on Bayesian confirmation. Clearly, using mathematical methods is one way to reduce philosophical complexity. But there are limits to this and I do not share the view that formal epistemology holds the key to answering basic philosophical problems. Perhaps, what makes philosophical problems distinctive (and maybe unanswerable) is precisely their complexity; their formalisation (based as it is on idealisation, simplification and abstraction) solves surrogates of these problems. For instance, I think there is an ineliminable role for explanation and explanatory considerations in ampliative reasoning—and this seems to resist formalisation precisely because there is more to explanation than satisfying an abstract or formal pattern. Or look at the prospect of understanding the role of framework principles in science using formal methods—it doesn't seem too good.

As I have noted already, issues in the ethics of science acquire urgency. One issue here is devising normative frameworks for the ethical conduct of research in science. It is plausible that there should be a core of ethical principles that ought to constrain scientific research. But then again, science is subject to cognitive aims too and, I take it, it is a very interesting question how norms of conduct in science can have a mixed (or double-sided) justification: cognitive and moral.

We need to pay more attention to the relations between central issues in the philosophy of science and topics and debates in other areas of philosophy (most notably, metaphysics, epistemology, and philosophy of language). Take the debates in epistemology about knowledge and evidence. These, to a large extent, are developed in isolation from what goes on in the philosophy of science. Conversely, debates about method and rationality in the philosophy of science have taken little notice of relevant developments in epistemology. This is hugely unfortunate. If philosophy of science cuts off its constitutive lore from the rest of philosophy, this will have disastrous consequences in the long run.

Finally, the complex landscape of the twentieth century philosophy should be re-drawn. The mushrooming of interest in the history of philosophy of science, the systematic attempts to re-evaluate and re-appraise the major philosophical schools and the major philosophers of science of the twentieth century are extremely welcome. The philosophical battlegrounds of the twentieth century saw many attacks on strawmen and a number of pyrrhic victories.

Selected Bibliography

Books

Knowing the Structure of Nature: Essays on Realism and Explanation. Palgrave-MacMillan, 2009.

The Routledge Companion to Philosophy of Science. Co-edited with M. Curd. Routledge, 2008.

Philosophy of Science A–Z. Edinburgh University Press, 2007.

Causation and Explanation. Acumen & McGill-Queens University, 2002.

Scientific Realism: How Science Tracks Truth. London & New York: Routledge, 1999.

Selected Articles

"Philosophy of Science." In D. Moran (ed.), *The Routledge Companion to the 20th Century Philosophy*, 618-657. Routledge, 2008.

"Past and Contemporary Perspectives on Explanation." In T. Kuipers (Volume ed.), D. M. Gabbay, P. Thagard, and J. Woods (Handbook eds.), *Handbook of the Philosophy of Science: Focal Issues*, 97-173. Elsevier BV, 2007.

"Putting a Bridle on Irrationality: An Appraisal of Van Fraassen's New Epistemology." In B. Monton (ed.), *Images of Empiricism*, 134-164. Oxford University Press, 2007.

"The Structure, the Whole Structure and Nothing but the Structure?" *Philosophy of Science.* 73: 560–570, 2006.

"What Do Powers Do When They Are Not Manifested?" *Philosophy and Phenomenological Research.* 72: 135-156, 2006.

"A Glimpse of the Secret Connexion: Harmonising Mechanisms with Counterfactuals." *Perspectives on Science.* 12: 288-319, 2004.

"Choosing the Realist Framework." *Synthese.* Forthcoming. DOI 10.1007/s11229-009-9606-9

"Is Structural Realism Possible?" *Philosophy of Science.* 68: S13-24, 2001.

"An Introduction to Carnap's 'Theoretical Concepts in Science.'" (together with the hitherto unpublished Carnap: "Theoretical Concepts in Science."), *Studies in History and Philosophy of Science.* 31: 151-72, 2000.

"On van Fraassen's Critique of Abductive Reasoning." *The Philosophical Quarterly.* 46: 31-47, 1996.

"Scientific Realism and the Pessimistic Induction." *Philosophy of Science.* 63: S306-314, 1996.

16

Joseph Rouse

Hedding Professor of Moral Science (Department of Philosophy), and Chair, Science in Society Program

Wesleyan University, USA

---

1. How were you initially drawn to philosophical issues regarding science?

As an undergraduate, I planned to study one of the physical sciences. In seeking to complete distribution requirements quickly, however, I took a philosophy course, and then another. Work in both philosophy and physics helped me realize that many of the questions drawing me to physics raised philosophical rather than scientific issues. After changing my major to philosophy, however, my scientific interests were temporarily side-tracked as I immersed myself in a new field.

I returned to philosophy of science in graduate school in the mid-1970s from an unexpected direction. I had planned to study 20th Century European philosophy with an interest in constructive engagement with the "analytic" tradition at a time when that was not fashionable. Immersing myself in Husserl, Heidegger and Merleau-Ponty, I quickly realized that philosophy of science was quite central to their work in ways not widely appreciated by their anglophone readers. I was also intrigued by the synergy between their work and the new post-empiricist developments in philosophy of science. Eventually I undertook a dissertation that approached then-current questions about relations between theory and observation, utilizing Heidegger's, Merleau-Ponty's and Samuel Todes's writings on equipment-use and perception.

In writing the dissertation, I gradually realized that familiar philosophical questions about the theory/observation relation were badly posed. Philosophers were rediscovering the recognition that scientific observation was "theory-laden," but had insufficiently grasped how both theory and observation were "practice-laden."

Theoretical understanding in scientific research outran philosophers' conceptions of theory, and experimental work far outran observation. Fortunately, I chose to dash off the dissertation relatively quickly, as if my original question were well-posed, and then accepted my first teaching post. Nearly a decade of further work, and some new developments in philosophy, were needed for me to work out an alternative approach in my first book, *Knowledge and Power*. Had I undertaken that reformulation as a graduate student, I would undoubtedly now be out of the profession.

Since then, I have repeatedly been drawn back into philosophy of science. Because I work on philosophy of science from within a more general philosophical concern to understand how bodily skill and language work together in intentionality, it was not a foregone conclusion that I would continue to work in philosophy of science. I remain in the field, because I keep seeing new possibilities for drawing upon more detailed study of the sciences to do broader philosophical work, and for that philosophical perspective to contribute to philosophy of science. Often, I have been led in new directions by encountering exciting work by others that takes on new significance in light of my somewhat idiosyncratic interests.

Not all of the work that has sustained my interest in philosophy of science was by philosophers. I was fortunate in the early 1980s to encounter the work of feminist scholars Evelyn Fox Keller and Donna Haraway, and sociologists Bruno Latour, Karin Knorr-Cetina, along with others working in Great Britain. Reading them at the same time that I was reading Ian Hacking and other philosophers looking at experimentation, and Nancy Cartwright's initial work on models and idealization, while also assimilating Sellars, Davidson, and Foucault, turned out to be extraordinarily productive for me. Around that same time, I started reading seriously in the history and philosophy of biology, initially to compensate for the disproportion of physics in most literature written for undergraduates. New work in philosophy of biology, and new conceptual issues emerging from developments in the life sciences themselves, continue to provoke my philosophical thinking. Later, Arthur Fine and Karen Barad encouraged me to consider the broader philosophical significance of the Bell inequalities and recent experimental work that stimulates the re-interpretation of quantum mechanics. The intersection of these issues with the revival of philosophical interest in causality has taken me in some illuminating new directions. Overall, I have continued to find intellectual sustenance triangulating among my broader philosophical concerns about intentionality and understanding, specific developments in the sciences and philosophy of science, and reflection upon scientific practice itself as a paradigmatic case of situated human understanding.

3. How has your work offered original contributions to discussion on science? What does your work reveal that others fail to appreciate?

The most central and distinctive theme of my work throughout my career in the philosophy of science has been a focus upon understanding scientific practice. A second theme has become increasingly distinctive as philosophy of science became more narrowly specialized and disconnected from work in other philosophical fields. My work in philosophy of science has been closely engaged with other philosophical work on metaphysics, epistemology and the philosophy of language and mind, and to a lesser extent with social and political philosophy. My attention to scientific practice has fundamentally influenced my conception of these broader themes, however, so I shall begin there. Attending to scientific practice has many far-reaching implications, which I should first disentangle.

The dominant tradition in the philosophy of science has been almost exclusively focused upon understanding scientific knowledge. The more restrictive American development of logical empiricism focused early attention upon understanding confirmation, explanation, the structure of theories, and the empirical content of theoretical concepts, all as part of a rational reconstruction of scientific knowledge and its justification. Their post-empiricist critics largely retained and even heightened the philosophical focus upon knowledge. Despite their historical interests, and their opposition to rationally reconstructive fictions, the early post-empiricists primarily addressed changes in theoretical beliefs, standards of justification for belief, and eventually the metaphysical commitments embedded in scientific theories taken as belief systems.

The retrospective focus of philosophy of science upon the structure of already-established theoretical systems oddly contrasts to the orientation of scientific researchers toward projects that look ahead to new research opportunities, and the possibilities for constructively transforming current understanding. A stark contrast between the retrospective orientation of philosophers and the prospective orientation of science over-simplifies the situation, however. Research scientists also read the recent history of their fields in ways that interestingly differ from philosophers' systematic reconstructions. From an interest in how the recent development of a field highlights new problems and opportunities for research, one acquires a very different sense of its history from that of a systematic reconstruction of beliefs. A research orientation focuses the field more narrowly, upon those issues that present significant and tractable research tasks or provide conceptual or experimental tools for taking up those tasks. Central to scientists' engaged reflection on their research fields is a topic that has been marginal to philosophical discussion: which questions and ways of addressing them are now scientifically significant? Undertaking work that is worth doing, and even important, is a crucial concern for scientific practice. For philosophers, the closest analogue to this concern has been an emphasis upon scientific explanation: scientific significance has thereby been effectively equated with explanatory power, which in turn has often been identified with the scope of laws and theories. While scientists are certainly concerned for their work to have wider rather than narrower implications, their judgments of scientific significance cannot be captured adequately in these terms.

Another important task of my work on scientific practice has been to develop a more adequate philosophical understanding of the concept of "practice." Epistemology has provided a ready-made (although not always particularly relevant or useful) set of concepts and issues for philosophical reflection upon scientific knowledge. Scientific practice has no comparable received view, although some misconceptions have had to be cleared away. Postempiricist philosophy of science, reinforced by eventual philosophical encounters with British sociology of scientific knowledge, encouraged a casual slippage from scientific practice to conceptions of the sciences as "social practices." Thomas Kuhn and other postempiricists seemed to ascribe important roles to scientific communities, typically understood by philosophers as defined by their supposedly shared beliefs, methodological commitments, and scientific values and norms. I have argued that conceptions of sciences as social practices, and the identification of practices with the activities of consensus communities, have been mistaken in multiple ways.

The biggest problem with conceptions of science as a social practice has been the implicit distinction between social relations among persons, and their encounters with the "natural" objects that the practice seeks to understand. I have argued that we must take seriously that science is a material practice, in which the construction of artificial and controlled settings, the extensive use of specialized apparatus, and manipulative interaction with the resulting phenomena, are integral to the practice. Moreover, the extension of scientific understanding beyond the laboratory (including establishing the extension of its concepts, in the semantic sense) often involves learning how to transform the situations in which the concepts are to apply in ways that make them more relevantly like the controlled settings of the laboratory or field site. Practices in the sense relevant to understanding science are not just what scientists do, but must be understood as incorporating the material and cultural settings of scientific work. Here my concerns have constructively meshed with the renewed philosophical attention to experiment, although most other philosophers of science have not emphasized how their work challenges conceptions of scientific communities, or of sciences as social practices.

I have also highlighted two additional difficulties raised by too-easy references to scientific communities. The first has been the ready assumption that scientific research communities can be defined by shared beliefs and values, even if only as shared presuppositions rather than explicit agreements in belief. Scientists often have quite different perspectives upon their field, and scientific "communities" can be riven by substantial disagreements among their members, even on quite central issues. What defines them as a research community need not be anything in common other than mostly overlapping mutual recognition as fellow participants in the field, and consequent interaction as interlocutors. Those interactions are often focused upon specific issues and stakes that define the field, although there need not be any common or agreed formulation of those issues and what is at stake in their resolution (often, practitioners disagree on how to formulate their disagreement, even while agreeing that they are talking about the same issues, and that they share a commitment to resolving those issues correctly). The second mistaken assumption has been that scientific communities are relatively self-contained, such that understanding scientific practice philosophically need not range further than the activities and beliefs of the relevant scientific community.

In situating scientific practices within a broader social and cultural context, I make common cause with many other scholars in science studies, especially those from anthropology, feminist science studies, and recent history of science. By the 1990s, many philosophers of science had begun to read British sociology of scientific knowledge. While almost entirely unsympathetic to the epistemological relativism and social constructivist anti-realism that was dominant in 1980s British sociology of science, philosophers of science have usually recognized the sociologists as addressing philosophical issues. That is not altogether surprising, given the extent to which the sociologists were influenced by Wittgenstein, Quine, Winch, and Kuhn. Most philosophers of science have found it much harder to recognize the relevance and significance of other work in science studies that rejects the assumptions most philosophers share with many of the sociologists. The anthropologists of science, for example, are not particularly concerned with issues of the justification of belief, for they are much more concerned with how scientific claims are understood as meaningful and significant. Scientists' grasp of what are important issues and what is at stake in their resolution also typically draws upon a much wider range of cultural resources than those that can be isolated within specifically scientific communities, even though there are obviously many concepts and practices that are distinctive to those communities. Yet even here those scientific developments sometimes take on a life of their own in other contexts, as scientific practices, tools and concepts gain wider circulation. Moreover, scientists are not isolated from these broader changes, which can affect the subsequent development of the science.

One striking contrast between the philosophers and the anthropologists of science concerns which fields of science are especially important to study. Someone whose acquaintance with the sciences was limited to the journals in philosophy of science might be forgiven for thinking that high-energy physics, general relativistic cosmology, foundations of quantum mechanics, evolutionary biology, molecular genetics, economics and cognitive science provided a near-exhaustive list of important scientific fields. Anthropologists are more likely to be found studying medical and psychological research, agricultural science, human biology (especially genetics, reproductive biology, and immunology), or nuclear power and weaponry. The philosophers emphasize science where it is most theoretically ambitious and philosophically contentious. The anthropologists focus instead upon where the sciences engage our everyday activities and self-understanding. Of course, an adequate understanding of science requires recognition of both.

The debates over scientific realism mark one especially prominent point of contrast between my work on scientific practice and the mainstream in the philosophy of science. I argued early on in *Knowledge and Power* and some earlier papers that we should reject the shared assumptions that identify scientific realism and empiricist, instrumentalist, or social constructivist anti-realisms as alternative answers to an important philosophical question about science. Commitment to taking seriously the issue of realism was of course one point in common between mainstream philosophy of science and constructivist sociology, and an utter lack of interest in this question within other approaches to science studies was one reason that these approaches have not engaged most philosophers of science. I am of course not the only philosopher of science to have argued against the very terms of the debates over realism, and against accepting any of the opposing alternative positions. Arthur Fine's contemporaneous defense of his Natural Ontological Attitude has been widely read and discussed within the field. Fine's and my responses to the realism debates have very similar direct consequences for the philosophy of science, but we differ in our argumentative strategies. Fine undertakes a minimalist strategy, arguing first that none of the standard realist or anti-realist programs are supported by good arguments, and second that the absence of any convincing arguments for these programs suggests that we can understand science perfectly well without taking a position on these issues. While I do not disagree with these claims, I think that we have a better understanding of why realist and anti-realist accounts of science are inadequate if we grasp why their inadequacy follows from an account of scientific understanding as situated within the practice of research.

This difference between Fine's and my strategies for circumventing the realism debates point toward the other major theme in my work, the effort to keep philosophy of science in closer conversation with broader currents in philosophy. Philosophy of science has become an increasingly isolated sub-field, both in the sense that it can be practiced without close attention to the primary developments in philosophy of language and mind, metaphysics, or social and political philosophy, and in the sense that philosophers in these "core" fields have not felt compelled to keep up with current work in the philosophy of science. That was certainly not the case in the heyday of Carnap and Hempel, and the work of their early critics from Quine and Sellars through Kuhn, Feyerabend and Putnam was also to integral subsequent developments in many areas of philosophy. Surely a major factor in the ensuing drift apart has been the recognition of the irreducible plurality of the sciences, and the expectation that philosophers of science would engage closely with detailed developments in specific sciences. Philosophy of science thereby became less accessible to other philosophers, and seemed to many to have less general import. At the same time, it became more difficult to expect philosophers of science to stay current with relevant work in a science and in philosophy of science, while also tracking developments in other increasingly technical fields of philosophy.

Yet the result has not actually been to isolate philosophy of science from other fields, but only to take for granted specific claims and approaches within those fields. Most philosophers of science have simply accepted without much question a representationalist semantics and philosophy of mind as the context in which they explicate scientific understanding. From the other direction, most metaphysicians, epistemologists, and philosophers of mind or language work against the background of a more or less explicit philosophy of science that underwrites their self-understanding in relation to philosophical naturalism. A significant part of my work has been to challenge entitlement to those presumptions on both sides. I have argued in some detail, drawing critically first upon the work of Davidson and Heidegger, and later of Robert Brandom and John Haugeland, that an adequate understanding of scientific work requires a broadly pragmatist and inferentialist conception of discursive practice and conceptual understanding, which challenges the standard versions of representationalism. These arguments have also been crucial to my criticism of the realism debates mentioned above. Yet part of my criticism of Heidegger, Davidson, McDowell, Brandom and Haugeland in turn has concerned their presumed philosophies of science, most notably in their conceptions of scientific laws and their neglect of scientific practice. Finally, in the other direction, I have argued that closer attention to scientific practice forces us to rethink what it would mean to be a naturalist in philosophy. A crucial desideratum for any philosophical naturalist is an adequate account of scientific practice and understanding in their preferred philosophical terms, and yet this task is rarely undertaken by self-described naturalists. I have argued in *How Scientific Practices Matter* that a very different philosophy of science (attentive to scientific practice) is required for a defensible philosophical naturalism.

One final theme that has been important to me, especially in my earlier work, has been to broaden philosophical conceptions of the political significance of the sciences. Philosophers tend to think of the politics of science as primarily concerning the acceptance of scientific theories, the politics of expertise, and the intersection of scientific theory with religion and political ideology. I have argued that we must understand the politics of science much more broadly, incorporating the ways in which the sciences materially change the world. Such changes happen on a smaller but crucial scale in laboratories, and take more dramatic effect as scientific practices, disciplines, and materials circulate outside the laboratory. One might take this point to be self-evident in an era of recognizably anthropogenic climate change, genetic engineering, pharmaceutical medicine, nuclear proliferation, and industrial synthetic organic chemistry, but philosophers of science have had remarkably less to say about the scientific transformation of the world we live in than about the scientific transformation of the beliefs of educated elites.

2. What, in your view, are the most interesting, important, or pressing problems in contemporary philosophy of science?

Not surprisingly, I find some of the issues central to my own concerns to offer some of the most interesting and important possibilities for further research. I will begin simply by mentioning two of the issues that I already raised in answering the previous question. Further attention to scientific practice in its many dimensions is surely one of the most significant opportunities for better philosophical understanding of the sciences and how they matter to us. Moreover, that work needs to be coupled with a constructive re-engagement of philosophy of science with other fields of philosophical work. Much work in philosophy of mind, language, metaphysics, and epistemology takes for granted philosophical conceptions of science that were current in the 1970s and 1980s, but have long since been surpassed by subsequent developments in the philosophy of science. Other work, especially in the tradition stemming from Sellars and Davidson, offers a relatively untapped resource for thinking about philosophy of science in new and illuminating ways.

Two other very broad areas of concern seem to me especially urgent. The first extends the past generation's move toward philosophical engagement with specific sciences. The second builds upon a nascent recent interest in understanding science in more complicated settings.

One obvious development within the sciences that deserves much greater philosophical attention is the scientific understanding of climate change. Global climate models themselves are an important and complicated example of theoretical modeling and simulation, which at least since the development of Monte Carlo simulations in the 1940s have helped blur the boundaries between experiments and theoretical models. The use of models in conjunction with new data-gathering to reconstruct paleo-climate patterns, and grasp the dynamics of climate fluctuation, is a kind of interdisciplinary scientific work that needs to be better understood. Of course, the importance of these models and the associated interdisciplinary syntheses in forecasting climate, and modeling the effects of changes in climate-related policy, should be obvious. One place where philosophical work might be especially instructive, however, is in trying to understand climate change in ways that acknowledge the more broad-band interactions between human activities and climate, especially in the light of efforts to mitigate greenhouse gas emissions and adapt to climate change. Current modeling mostly presumes that anthropogenic greenhouse gas emissions are a relatively simple interface between climate as a physical system, and human life as an economic and political system. As modeling becomes more fine-grained, it will likely need to move beyond that presumption, in ways that will be especially important because it challenges the autonomy of scientific disciplinary perspectives.[1]

Another topic that is already gaining considerable philosophical attention, and deserves more, are the various efforts to reintegrate evolutionary and developmental biology and ecology, in evolutionary-developmental biology (developmental evolution), ecological-developmental biology, and developmental systems theory. These new and rapidly changing fields also engage with the nascent disciplinary shift from genetics to genomics and proteomics, which have the potential for dramatically transforming fundamental concepts and research orientations in the life sciences, and in broader understanding of life and evolution. Yet a third important area that is already attracting philosophical attention is the interaction of neuroscience, psychology, cognitive science and even phenomenology in developing enactive accounts of perception and cognition. Generally, attention to new areas of inter-

---

[1] Thanks to Emily Rowan, whose undergraduate thesis research in the Science in Society Program at Wesleyan University helped convince me of the importance and difficulty of this topic.

disciplinary and especially conceptually innovative or revisionist scientific research is a promising locus of research possibilities for philosophy of science.

The second major area of concern brings together disparate topics under the rubric of "science in the real world." One important segment of this domain is a shift in philosophical attention from laws and theoretical unification to the analysis of causation in complex situations. Increasingly, philosophers recognize that causation is not a univocal notion, and that complex causal systems cannot always be understood by reducing them to their component interactions. Moreover, there are complicated epistemic challenges in how and when to infer from correlation to causation, and how to assess the adequacy of our causal claims. Alongside this anti-Humean turn, there is also a growing emphasis upon heuristics, approximation, error-analysis and error-reduction, idealization, and the role of practical engineering methods and practices in supposedly "pure" science.

Yet a shift to understanding "real-world" science also acknowledges the importance of social institutions and contexts. The institutions and practices of scientific disciplines, and interdisciplinary fields shape the directions of research. Government funding and other sources of patronage or sponsorship, from corporations to foundations, importantly affect those directions. Universities, publishers, commercial suppliers of instruments and materials, and various non-governmental organizations are also integral to scientific practice, and call for increased philosophical understanding. The importance of scientific work in shaping many aspects of public policy further complicates philosophical understanding of science as it is actually practiced. The political significance of the sciences is heightened the more we recognize them to be material, world-changing activities, and not merely the basis for changes in belief. On the one hand, we need to understand how scientific expertise should contribute to democratic assessment of policy, including scientific research priorities. On the other hand, the rise of evidence-based medicine and other strategies for assessing the application of knowledge requires new critical models in philosophy of science. What constitutes appropriate evidence for scientific belief-formation may not play the same role for policy analysis. The needs and concerns of disparate stakeholders can appropriately shift the thresholds of acceptable evidence when there are substantive consequences to the resulting decisions. Many traditional philosophical conceptions of science look increasingly abstract and thin in light of its material and institutional complexity.

5. **Where do you see the field of philosophy of science to be headed? What are the prospects for progress regarding the issues you take to be most important?** *and* **4. What is the relation between philosophy of science and scientific practice, science policy, or efforts for social justice? Can there be a more productive relation? Is this desirable?**

Philosophical attention to scientific practice has promising prospects. Much good work is now being undertaken on modeling, experimentation, and conceptual articulation, and there is increasing recognition that scientific practice is a distinct topic that calls for new philosophical approaches and conceptions. The formation of the international Society for the Philosophy of Science in Practice both reflects this growth in philosophical interest, and provides a venue to sustain ongoing work. This work on scientific practice has been accompanied by extensive attention to causality, and growing interest in standards of evidence for use in policy-making or other practical decision contexts. There has also been some new work on the place of science in a democratic society, and some localized discussions of specific issues in science policy (e.g., concerning climate change). These latter topics are currently less far along substantively, in large part due to reliance upon models of science that are inadequate for these purposes. That situation may well change, however, as new work on scientific practice is integrated within studies of the broader political significance of the sciences. After all, when first broaching a relatively new and controversial topic, there is some rationale for reliance upon relatively conservative philosophical models; as the topic becomes more widely recognized as relevant to the field, that rationale will be less compelling, and criticism of the extant approaches will also likely broaden the discussion.

There is even greater reason for confidence that philosophers of science will engage effectively with new (or newly prominent) issues in the sciences. Serious engagement with the sciences is now central to the ethos of the field, and philosophers are readily drawn toward new research opportunities stemming from new scientific developments.

I have much less confidence in the imminent prospects for work that more effectively re-connects philosophy of science to issues in philosophy generally. The biggest difficulty here arises from institutional barriers. The need for substantial engagement with the sciences makes it increasingly difficult for philosophers of science also to maintain sufficient background and involvement in other areas of philosophy to make credible contributions there. By the same token, the technical complexity of much work in philosophy of science makes it less accessible to other philosophers. These institutional barriers are reinforced by the lack of an adequate vocabulary for connecting philosophical work on theoretical models, experimental practice, or conceptual development to current issues in philosophy of mind, epistemology or metaphysics. The latter fields still remain closely tied to older ways of thinking about the sciences, and may not be so readily adaptable to issues arising in the philosophy of scientific practice. A change in that situation may only result if a crisis in these other philosophical fields were to encourage renewed attention to recent work in the philosophy of science as a promising source of novel insights.

It is more difficult to assess whether more sustained philosophical discussions of science and social justice will be forthcoming. There have been occasionally promising forays into these issues, but they have largely been driven by the recognition within social and political philosophy that the sciences matter to the prospects for social justice. I suspect that the extent of philosophical attention to these concerns will not be determined by other developments within the philosophy of science.

Selected Bibliography

Books

Knowledge and Power: Toward a Political Philosophy of Science. Cornell University Press, 1987; Japanese translation, 2000; Chinese translation, 2003.

Engaging Science: How to Understand its Practices Philosophically. Cornell University Press, 1996.

How Scientific Practices Matter: Reclaiming Philosophical Naturalism. University of Chicago Press, 2002.

Selected Articles

"Laboratory Fictions." In M. Suárez (ed.), *Fiction in Science: Philosophical Essays on Modeling and Idealization.* New York: Routledge, 2008.

"Naturalism and Scientific Practices: A Concluding Scientific Postscript." In C. Michael Mi and R.-L. Chen (eds.), *Naturalized Epistemology and Philosophy of Science*, 61-86. Amsterdam: Rodopi, 2007.

"Practice Theory." In S. Turner and M. Risjord (ed.), *Handbook of the Philosophy of Science. Vol. 15: Philosophy of Anthropology and Sociology*, 630-681. Dordrecht: Elsevier, 2007.

"Epistemological Derangement." *Studies in History and Philosophy of Science*. 36: 835-847, 2005.

"Heidegger's Philosophy of Science." In H. Dreyfus and M. Wrathall (ed.), *A Companion to Heidegger*, 173-189. Blackwell, 2005.

"Merleau-Ponty and the Existential Conception of Science." In T. Carman and M. Hansen (ed.), *The Cambridge Companion to Merleau-Ponty*, 265-290. Cambridge University Press, 2004.

"Kuhn's Philosophy of Scientific Practice." In T. Nickles (ed.), *Thomas Kuhn*, 101-21. Cambridge: Cambridge University Press, 2002.

"Vampires: Social Constructivism, Realism, and Other Philosophical Undead." *History and Theory*. 41: 60-78, 2002.

"Beyond Epistemic Sovereignty." In P. Galison and D. Stump (ed.), *The Disunity of Science: Boundaries, Contexts, and Power*, 398-416. Stanford: Stanford University Press, 1996.

"Power/Knowledge." In Gary Gutting (ed.), *The Cambridge Companion to Foucault*, 92-114. Cambridge: Cambridge University Press, 1994; revised for Second Edition, 2005.

17
Patrick Suppes

Lucie Stern Professor of Philosophy, Emeritus
Stanford University, USA

---

1. How were you initially drawn to philosophical issues regarding science?

I will give a rather diffuse answer by beginning quite early. As I remember, perhaps somewhat distortedly, my thinking about any kind of conceptual issue in science, or in other matters that had any kind of philosophical edge, first began in what was then called junior high school. Born in 1922, I entered the seventh grade, the first year of junior high school, in September of 1933 at the age of 11, having skipped one year of elementary school. The importance of junior high school was not really the shift from one school building to another, but the dramatic movement into a special program for gifted students in the public schools of Tulsa, Oklahoma. The world opened up in front of me rapidly, as we were encouraged to talk about anything on every occasion, especially by the nourishing and encouraging teacher of social sciences. She had us reading whatever we could get our hands on about politics, culture, and the economic problems of the Great Depression. This was right up my alley. I quickly found I liked having an opinion on everything and was quite anxious to express myself in class discussions. I would mark this class as the real starting point of my having, in a way I did not fully understand at that time, a philosophical interest.

The second formative development also began during these junior-high-school years: my skeptical withdrawal from the belief in the doctrines of Christian Science as set forth by Mary Baker Eddy. I had since the early years of elementary school attended Christian Science Sunday School, but it was only at the age of 12 in the eighth grade that considerable skepticism rapidly developed about her outlandish philosophical views, which were developed in the heyday of philosophical idealism in the 19th century. Most surely, I and my fellow young skeptics could not believe that her account of disease and its purely mental state could possibly be correct. This was, in fact, perhaps the most important direct conceptual conflict that facilitated my developing interest in philosophical matters. There seemed to be, in the context I was in, an inevitable and dramatic conflict between science and religion, and in my case, science carried the day.

In my high-school years, running from September 1936 to June 1939, the program for gifted students continued, and we were encouraged to read widely and talk extensively about our interests. Writing now just from memory, the book I remember as stimulating the most thought on my part was Bertrand Russell's popular book on Einstein's theory of special relativity. I can't remember the puzzles it provoked in my mind, but I certainly know I was puzzled, and realized that I did not in any serious way understand the physics and the concepts of space and time that were being discussed. That surface glitter of clarity, so characteristic of Russell's writing, seduced me, as it has many others on many other philosophical topics, but at the same time did not provide a very deep understanding of special relativity. Yet I got what was important—a sense that all was not simple and clear and well understood about space and time, and that was perhaps my first glimmering of what it might mean to have a philosophical interest in physics and its account of the physical world. I learned little of real importance in my high-school physics class. (The instructor gave an impression, probably correctly, of being untrained and uncertain about much of the physics he was supposed to teach.)

As an undergraduate, I first took an introductory course in philosophy as a sophomore at the University of Chicago in the academic year 1940-1941. The course was taught by Marjorie Green, but she had also a number of guest lecturers. I was intrigued by this course, which began with Ancient Greek philosophy, and I was especially drawn to the brilliant lectures on Aristotle by Richard McKeon, who, I found out later, was one of the major influences in the development of the humanities at the University of Chicago under the regime of President Robert Hutchins. McKeon, by the way, was the kind of philosopher—in his case, even more a historian of philosophy—who gave brilliant and what seemed to be wonderfully clear lectures, and yet wrote in a way that was dense and often opaque. As some of us who took a look at his book on Spinoza in those days agreed, he was much more difficult to read than Spinoza himself. There was not much in this introductory course that stimulated a direct interest in the philosophy of science, but it deepened considerably my interest in philosophy.

There is also one other event from my undergraduate days, quite different from what I have said about Chicago. This was when I was spending a year at the University of Tulsa. I can remember having extensive, tortured, and often obscure arguments and discussions about the nature of infinitesimals in the beginning course in ordinary differential equations I was taking. The lecturer himself did not, in retrospect, have a very good understanding of these matters from a rigorous standpoint, but he was interested in the problem of giving a clear explanation of differentials and recognized there were conceptual difficulties. I certainly did not come to any really deep or clear understanding, but again what was important was that I recognized there was a problem and got a feeling for how subtle the solution of such a problem might be. Also during this period (1940), just before being called into the army, I read at the University of Tulsa on a regular basis various reviews in the *Journal of Philosophy* by Ernest Nagel. These were, of course, directly about matters in the philosophy of science, and I liked very much the way he analyzed and criticized work in the philosophy of science, even when I did not fully understand all the issues involved.

I spent my fourth year as an undergraduate again at the University of Chicago, but now in uniform as an army cadet being trained in meteorology. Because my undergraduate work was mainly in mathematics and physics, I qualified for this program and underwent from November 1943 until graduation in September 1944 an intense course of the study of meteorology as a science to prepare me, at least in part, to be a weather forecaster for the Army Air Force. Not much direct philosophical thinking was generated by the course of meteorology, but I was excited by the beauty of the one graduate course in hydrodynamics I took as part of this training. I still have my notes from that course, and remember well the broad outlines of the subject. It also was, as meteorology was as well, a warning as to how much more complicated real physical phenomena are than the elementary examples I learned earlier in my first courses in physics at the University of Tulsa in my junior year.

Immediately upon graduation, with a bachelor's degree and, at the same time, commissioned as a second lieutenant in the Army Air Force, I was sent to the Army Air Force base in Salt

Lake City for shipment overseas. I finally arrived early in 1943 in the Solomon Islands, just as Guadalcanal was being secured, and moved further north, first to one of the small Treasury Islands just south of the large island of Bougainville, which was held by the Japanese until the end of the war, and then after a few months for almost a year to one of the Green Islands, which are obscure, also small, but strategically placed for the war at that time, south of the Truk Islands and close enough to Rabaul, New Britain to the west to run massive bombing raids continually both to Rabaul and to Truk. In the isolated environment of the Solomon Islands, I really began to read philosophy in a serious way, using as a guide the systematic lectures I had heard at Chicago two years earlier, starting with Aristotle. I still have my McKeon edition of selected works of Aristotle that I carried with me throughout the South Pacific for nearly three years. I also began reading Kant, but Aristotle was much more my focus. I also read some Aquinas, but as I remember, did not have a volume of his writings in the South Pacific. I did have the big volume of Aristotle and Kant's *Critique of Pure Reason*. I do not want to give the wrong impression about carrying these books around. I read them in a desultory and scattered fashion, no doubt, and my understanding of them was certainly only partial. But they did the important thing—they whetted my appetite for a deeper understanding of philosophy and its relation to science.

After the war, I spent 1946 deciding what to do, and finally, after applying a few months earlier, entered graduate school in philosophy at Columbia in January of 1947. In the spring semester that year, the course I still remember well was the seminar on the logic of F.H. Bradley and John Dewey given by Ernest Nagel. I was attracted at once to Nagel's patient but critical assessment of these two original but deeply flawed thinkers, and it was Nagel who became my mentor and introduced me to the philosophy of science, which quickly became my focus as an academic subject. Once under Nagel's influence, I was on my way to becoming serious about philosophy, receiving my Ph.D. in June of 1950.

2. What, in your view, are the most interesting, important, or pressing problems in contemporary philosophy of science? *and* **5. Where do you see the field of philosophy of science to be headed? What are the prospects for progress regarding the issues you take to be most important?**

I have listed what I consider important problems in the categories of individual sciences at a fairly general level. My remarks will emphasize physics and neuroscience, but let me start with statistics. The controversy about the foundations of probability, so intense a couple of decades ago in the conflict between subjectivists and objectivists, or in other terms, between subjectivists and behaviorists of the Neyman-Pearson stripe, is now not so intense, and the center of statistics has moved to the pressing and complicated problems of large-scale data analysis. But the foundational problems are still there, and it is easy to enumerate various aspects of them. They have, however, been so much discussed that I will only emphasize, as I have previously, the importance of continued conceptual analysis of the nature of randomness. I like to make the distinction between random sequences of events and mechanisms of producing such sequences. Much of the technical literature is concentrated on the analysis of sequences of outcomes and what criteria they must satisfy observationally in order to be regarded as random sequences. From a general conceptual point, it is also of great interest to try to understand what kinds of mechanisms, especially physical mechanisms, are capable of producing randomness. The first problem is now very much a technical problem centered primarily in complexity theory, but the second remains as a general part of the foundations of statistics. There are, of course, a number of mechanisms in quantum mechanics that should produce random sequences, and most people believe that if thoroughly tested they would, but the entire matter is far from a settled one and certainly warrants further intense philosophical and scientific discussion.

I turn now to psychology, where there are, of course, endless conflicting conceptions that need philosophical analysis. In the present framework I will just mention a set of related problems that have been very much on my mind in my own recent thinking about philosophy and psychology. These are problems that center around norms. First, from a psychological standpoint, what is the right theoretical framework to analyze the empirical status of norms? Is it an appropriate aspect of fundamental psychological theories of behavior and cognition to be concerned with the empirical status of norms? My own answer is clearly affirmative, and, in fact, I am critical of purely philosophical attempts to develop anything like a complete theory of norms. I think that any adequate theory of norms must deal in the intricate ways in which norms are developed, modified and sometimes rejected. The details of these processes are matters to be studied scientifically, not to be settled by general philosophical argument. A particularly pressing problem is how to view the problem of conflict of norms. There is too much happy thought in philosophy that seems to assume that, in given problem domains, there is no conflict of norms. My own view is that conflict of norms is everywhere, and is the most severe direct limitation on too general conceptions of rationality in behavior or in cognition. This is not the place to lay out the case for my claim about the ubiquitous character of norm conflicts, but it is my own way of criticizing any attempts to use either theoretical models or those based on practical reasoning to claim that pure deliberation ever reaches practical conclusions leading to complex actions. In almost all substantive cases, norms are in conflict in decisions related to actions, and the problem of how to handle these conflicts is central to philosophical conceptions of rationality and psychological theories of decision-making. The simplest and most beautiful examples of such conflicts in norms about decision are to be found in game theory and in the traditions of welfare economics, first really well-exemplified in Kenneth Arrow's well-known book *Social Choice and Individual Values* (1951).

Turning to closely related problems in economics, I also restrict my thoughts to the development of a deeper theory of rational choice. Economists are faced with the problems I just mentioned for philosophers and psychologists of the clash of norms, and in fact, the best examples I just mentioned, and similar ones in game theory, have been to a large extent developed within the general framework of economic thinking. But I want to criticize economists' theory of choice from another standpoint, and this is the shallowness from a psychological perspective of the theory of preference, whether we are discussing the empirical matters of revealed preference or the rational theory of how preferences should be structured for a rational agent. Preference itself is a useful concept, and has done much work in modern theories of choice, but it is also a psychologically superficial concept, in the sense that it is not derived ordinarily from deeper, more general theories of behavior and cognition. Also, for the reasons I have stated above in discussing matters of norms, I am skeptical of too much of an emphasis on optimization and not enough in economics on the natural and central place of habits and the free associations that take over when habits have completed their role. Choices are not usually uniquely dictated by habit. An example I have used on several occasions is that of my own preference for wine over beer when ordering at a restaurant. The habit is not for some particular vineyard or vintage, but just for the sharp decision of wine, not beer, for me. After that, the idiosyncrasies of the wine lists of individual restaurants take over, and my free associations work through that list. I end up, hit-and-miss, ordering something, perhaps constrained by the price and the vintage, but constrained no more. Free associations complete the final, distinctive choice of the particular bottle purchased. So I am, in the end, with Hume on the theory of associations as the most fundamental psychological theory of the mind. (For elaboration, see Suppes, 2003.) So as not to be misunderstood in this remark, I emphasize that we now know what Hume did not. From a computational standpoint, associative networks can be as powerful as universal Turing machines.

Let me now turn to physics. In terms of my own knowledge and interests, I know most about quantum mechanical entanglement, and think that the mysteries that still surround entanglement will continue for some time to bedevil philosophical thinking about physics and the nature of the physical world. It may turn out that things are much stranger than we thought, just on the basis of entanglement considerations alone. But equally mysterious and difficult to get to the bottom of is the current proposal in astrophysics of multiple universes. The idea that we live in only one of a multitude of universes is one of the more mind-boggling concepts introduced in theoretical physics in the last several hundred years. It is not that the general problem is new; it is closely related, in my own judgment, to the old problem of whether the universe is eternal or has a fixed beginning, a problem discussed in ancient times, judiciously analyzed by Aquinas, with the claim that whether the universe was eternal or had a fixed beginning is a matter that must be left to faith and not to reason. It also forms a principle antimony in Kant's short list in the *Critique of Pure Reason*. The multi-universe hypothesis is, in many ways, a natural extension of these earlier puzzles, but breathtaking because of the new technical interest in the concept and the possibility of this being the true state of affairs. It is easy to make this list much longer, for example by referring to the continuing mysteries of dark energy and dark matter, or some of the new theories of space and time. I think from a broad philosophical standpoint, the general importance is that our view of the fundamental aspects of our physical universe are now in a very unsettled state. The calm of many earlier centuries about these matters is far from prevailing today. Philosophers of physics should not be out of work any time in this century.

Finally, I turn to neuroscience and a problem that is also a physical problem. This is the fundamental one of how human brains compute. What we know at present about brain computations is severely limited. There are many conjectures that seem sound in their general formulation; for example, the hypothesis that a network of weakly coupled oscillators, synchronized by phase-locking, is a model that will solve a fair number of problems. These oscillator-type proposals use one of the most widespread models in classical and quantum physics. They certainly seem promising, but we are very far from having anything like a thoroughly understood theory of such oscillator computations as being central to brain processing.

On the other hand, I conclude as unreasonable, on the basis of current evidence, the claim by some that the brain must be some kind of quantum computer. The possibilities for classical computations are much too rich and as yet too unexplored to think that the unstable, rapidly decohering properties of quantum entanglement form the basis of brain computations in the rather rich and noisy environment of our brains. But there are many problems of getting the observations and having the right detailed models, to analyze the observations in order to conclude that current oscillator ideas are close to being correct. As in the case of multiple universes, but perhaps not quite so esoteric, there are bound to be some great surprises as we come to a full understanding of brain computations. Reaching a much deeper understanding than we now have will not only change our conception of the human mind, but also our general conception of computation, and will undoubtedly cause waves in more than one direction of philosophical thought.

3. How has your work offered original contributions to discussion on science? What does your work reveal that others fail to appreciate?

A good many scientists, especially towards the end of their careers, turned to the philosophy of science. My own career has been the reverse of this. I started out, and was trained, as a philosopher of science: my Ph.D. was in the Department of Philosophy at Columbia University. But not too long after I began teaching at Stanford as a faculty member, my scientific interests occupied at least as much time as my work in the philosophy of science, and, for many years, probably more. So, I have been discussing and working at the interface between science and philosophy of science for many decades. Also, I have worked in a number of different fields, and naturally I have ideas about work of mine that has not been, I think, fully appreciated, but I also am now old enough to understand how complicated the process is of any particular new idea being appreciated. Some good ideas do not get much appreciation for many years, and some bad ones get too much attention from the very beginning. But I am not complaining about that. It may well be that some of my worst ideas got more attention than they deserved. I just think it is too difficult to comment in an objective way on this matter of appreciation; in fact, to put the matter in a summery fashion, I am reasonably satisfied with the attention that most of my work has received. I do not feel that I have been underappreciated.

On the other hand, it is perhaps useful, in a more general answer to this question, to note that the general appreciation of science has declined in philosophy over the course of my own career. I can remember when it was felt that some kind of view of theories of space and time were part of the general education of any philosopher. Now this is not the case. The problem is solved by having in a serious department of philosophy at least one person who knows something about the foundations of physics. The rest of the department is left scot-free. But this is only part of the general trend. In the past several decades philosophy has, it seems to me, moved away from science. This will probably change, and in some respects is already changing in the following way. The most general change in philosophy since I began teaching in 1950 is the move from the central position occupied by epistemology, philosophy of mind, philosophy of language, and, to a lesser extent, history of philosophy, to a much greater emphasis on ethics and moral philosophy. What is beginning to happen is that work in ethics and moral philosophy is becoming much more involved in the social sciences, especially in economics and psychology, whose concepts and results often have a direct bearing on problems considered central to moral philosophy. This is a happy event in my own view, and one that will tend to bring philosophy and science closer together than has been the case for some time.

I also make the optimistic prediction that the philosophy of mind will be much more influenced by psychology and neuroscience in the next few decades than it has been in the past half-century. I am of course biased in this matter because of my current interest in neuroscience and the experimental study of brain computations, but in any case the brain will be a great focus of much scientific work in this century, and it is inevitable that some of the most significant aspects will spill over into philosophy.

4. What is the relation between philosophy of science and scientific practice, science policy, or efforts for social justice? Can there be a more productive relation? Is this desirable?

I have already stressed my interest in and concern for the development of closer attention on the part of philosophers of science to scientific practice. It is good to see that evermore sophisticated analysis of scientific practice, especially experimentation, is the focus of a good number of younger philosophers of science. A difficult example of the kind of work we need is to be found in the study of instrumentation in ancient astronomy. Such a great historian of ancient astronomy as Neugebauer said that he did not feel equipped to make a detailed study of this subject. There have been useful efforts, but it is still a subject that needs much more detailed attention. There have also been good recent studies of instrumentation in more recent science, for example, the 19th century. Such efforts are much too be commended and reflected upon in terms of their philosophical implications. It is often claimed, and I think rightly, that many important breakthroughs in physics have been almost entirely dependent on the development of new and better instrumentation. It is not at all impossible to believe this; what is needed is more detailed and nuanced accounts of how these took place, and philosophers of science could certainly play a useful role in analyzing both the nature of such events and their consequences.

I turn now to some philosophical opportunities concerning science policy, drawn from Suppes (1982) and Suppes (1984). I mention five broad categories of problems that certainly could benefit from deeper study: numerical models, statistical methodology, philosophy of applications, analysis of predictions, and distributive justice and resource allocation for scientific research.

Numerical Models

The explicit relations between theory, mathematical model, and computation are often only briefly and incompletely laid out in many policy frameworks. There are several aspects of these relationships that need further analysis. First, it is characteristic of the mathematical models widely used in policy studies not to have solutions in closed symbolic form for many of the cases of interest. Linear programming applications are typical. Suppose that we are studying an economic or logistical problem involving several hundred variables or possibly several thousand. There is no hope of having a symbolic solution. We obtain solutions for particular numerical sets of coefficients. Moreover, in all but the most trivial cases, we will not have a unique solution but a convex polyhedron of solutions. There is a conceptual shift here from the ideals at least of classical science and its search for unique closed-form solutions. After all, it is an old methodological axiom of physics that when the initial and boundary conditions are completely specified the behavior of a physical system is uniquely determined. This is not at all the situation in a wide variety of the mathematical models used in policy studies. To the charge that a real watershed change is taking place in moving away from closed-form solutions to numerical solutions, the response can be made that this is the case already for much of modern physics and quantum chemistry, but still it has remained of interest in physics and in chemistry to compute solutions for trivial one-dimensional cases, etc., in order to give a conceptual sense of theoretical developments. In contrast, in the kinds of applications I am referring to here in policy studies, such small-dimensional studies are of really no interest at all. In this respect, it might well be said appropriately that mathematical models are used in policy studies much more in the style of engineering than in that of pure science.

But I do not want to overdo this point either. Already in the classical gravitational case of the three-body problem, closed-form solutions are not possible. It was shown by Bruns now almost a hundred years ago (1887) that in the three-body problem the ten conditions of conservation are the only algebraic constraints possible—six equations coming from the conservation of momentum, three from the conservation of angular momentum, and one from the conservation of energy. More fundamentally, infinite series approximations are shown to be essential by Poincaré. Indeed perhaps Poincaré's most celebrated theorem in celestial mechanics is that if the infinite series used converge at all, they cannot converge uniformly for all values of time and simultaneously for all values of the constants, even when held between fixed limits. It is not to the purpose here to pursue the three-body problem in detail. What is important is that the kinds of purely numerical approaches characteristic of the mathematical models used in policy studies are also characteristic of much advanced science. We have not made as much of that in the philosophy of science, in my judgment, as we should have. In fact, there seems to be a general philosophical distaste for analyzing the specific meaning of real numerical data.

Statistical Methodology
In policy studies that use the kinds of models I have been talking about and that appeal to real data, there often needs to be a deeper development of the statistical methodology. It is a point of some interest, from the standpoint of the philosophy of science, to analyze and criticize the relation between such developed mathematical models and statistical aspects of the data or parameters being used. I very much agree with that famous remark of de Finetti's that in building a house it is better to build on sand than on the void, but we like to know as much as we can about foundational matters. A very good example of a problem that occurs repeatedly is the desirability of knowing something about the joint distribution of parameters. All too often we are content to look only at their marginal distributions and ignore evident data on dependencies. For philosophers, such questions are of particular interest because of the close connections between causality and evidence of probabilistic dependence. Notice that this is not so much a criticism of the general statistical methodology applied but rather of the way in which the data that are available are used and interpreted.

Philosophy of Applications
In the sense of ancient Greek mathematics, the mathematical models of linear programming and the like are justified procedures, in the sense that the constructions made are mathematically correct solutions for the given constraining conditions. But even at the formal level, interesting and important questions remain. It is characteristic of the kinds of applications I have been discussing that computational algorithms dominate the mathematical scene, especially when the initial boundary conditions, as in linear-programming problems, are expressed by sets of linear inequalities. It is easy enough to talk about the form of these constraints. It is quite another matter to seriously discuss the constructive solution of a minimization problem subject to these constraints. This is a kind of highly finitistic constructivism that does not yet seem to have a proper place in our thinking about the philosophy of mathematics. Here constructivism is not a broad philosophical issue but a focused one of practical consequence. It is also often unobvious why extremely good algorithms work as

well as they do. A good example of this is the subtle analysis of the simplex method of solving linear programming problems by Stephen Smale (1981). But what is even more needed is an explicit philosophy of application. Philosophers of various special sciences, but especially physics, have spent some time talking about the nature of experiments and the character of experimental data. Not as much has been done in this direction in a modern spirit as seems desirable, but far less has been done in analyzing the nature of the data used in the application of mathematical models in policy studies in order to determine some optimum course of action. A fair amount of romantic numerology may be found in some important parts of science, cosmology being perhaps the best extended example, but it is very much my impression that the real romance of numbers in the modern world is to be found in the discussion of data to be used in the determination of policy. This is to be seen in everything from elementary and off-the-cuff computations in Congressional testimony on new tax legislation, concerned with the impact on revenues of tax changes, to some of the splendid fables that are spun by nuclear engineers estimating the probability of core meltdown in a nuclear plant.

Analysis of Predictions
Quantitative determination of optimum policies is in spirit closely related to problems of quantitative prediction, which is not all of science, and, in fact, it might be claimed, a rather small part of it, but it represents an important and essential element. In some few areas there have been real efforts to make a serious study of predictions. The daily news is full of evaluations of how poor various economic prognostications are, or have been. Perhaps some of the best examples are to be found in the extended study of weather forecasting, either by numerical models or by individual forecasters. A careful and detailed study, surveying the earlier literature as well, is that of Murphy and Winkler (1984). It is hard to find an analogue to the Murphy and Winkler study of probability forecasting in meteorology in corresponding studies of policy recommendations. There is one area in which the absence of such data is particularly disappointing. As George Dantzig liked to emphasize, linear programming has become more popular in industry than anyone dreamed of thirty years ago. But it is not only the particular methods of linear programming. There is a wide array of methods available to modern bureaucracies and businesses to be used in determining policy, or I should say policies, for

the methods range across pricing policy, inventory policy, profit-maximization policy, medical-benefits policy, and compensation policy, to mention a few. What is almost totally missing from the literature is detailed factual follow-up studies of how well policy implementations worked, compared to the forecasted results. It is, it seems to me, the most serious intellectual weakness of the modern government or business executive not to be very much on top of this question. He or she needs to have a skeptical, and at the same time, informed view of how good or bad his quantitative problem solvers are doing. It might be claimed that the information is not available because it is proprietary, but the almost total absence of good data on the point I am discussing shows that the neglect runs much deeper.

No doubt the rapid development of information technology and the computational power of even relatively inexpensive computers have made the use of systematic predictive methods a widespread reality. But the use of such methods for policy determination, whether by a government department or a department store, is no panacea. In the past, policy was often made in such an intuitive and personalized way that little hope of systematic methodological discussion seemed possible. Now the opportunities are unbounded. For philosophers who find the study of past, or even present, pure science too quiet and serene an activity, the arena of policy studies may have just the right mix of sophisticated concepts and raw emotional furies.

Distributive Justice and Resource Allocation for Scientific Research

A variety of literature in economics and philosophy makes the point that issues of allocation must ultimately be judged by criteria of distributive justice. It is worth examining what can be said about distributive justice in the special case of allocation of resources for scientific research. Two different questions naturally arise. The first concerns the allocation to individual disciplines, given broad governmental or societal agreement on the total to be allocated, and the second concerns the allocation to research in competition with other kinds of demand.

Before considering either of these questions, however, there is one reservation that must be made explicit. We can take as an approach to distributive justice in the allocation of resources the use of market forces to make the allocation. To a large extent this is obviously not the case anywhere in the world in terms of basic scientific research. But when applied research as well as basic research is considered, then there is a decentralized allocation, partly to be accounted for by market forces, in making the allocation to applied research and occasionally to basic research conducted by private enterprises, especially large corporations, throughout the world. Even within governmental allocation there can be decentralization, because allocations by one government agency can be made in independence of and in ignorance of the allocation made by another one. I think that the play of such market forces and, even more important, the encouragement of decentralization are factors in the allocations that are actually made that should not be ignored and, in many cases, should be encouraged. All the same, I want to concentrate on appraisal of the total allocation by whatever mechanisms it is made. We can, if we want, think of this appraisal being made from the standpoint of a policy committee, not itself concerned with the particular mechanisms of allocation, but with judging the appropriateness of the total allocation made by various means, both private and governmental, and by various instrumentalities.

With respect to both these questions, there are some second-order considerations of justice that are not often discussed, and yet are of great practical importance in working out actual allocations among disciplines. One is a consideration of continuity. Many would consider it irrational simply to change the allocations in a highly discontinuous way from one year to another because, perhaps, of changes in the judges asked to make the evaluations. Smooth transitions from one allocation to the next, from a practical standpoint, seem essential to satisfy a criterion of justice that is not well articulated in current theories but that comes under the Aristotelian criterion of consistency of judgment.

When we turn to the competitive allocation, as a whole, to research versus other societal demands—for example, for education, for health, or for welfare—it might seem that criteria of distributive justice could play a central role, but I am skeptical of any explicit application of current theories of justice to this allocation. As in the matter of internal allocation, criteria of justice can be brought to bear in thinking about and evaluating detailed considerations, but it seems to me that, except for weak constraints as exemplified in Pareto-type principles, the judgments of the suitability of individual dimensions of evaluation will depend upon the sophisticated intuitions of experienced policymakers. Moreover, I am skeptical that these intuitions can be made theoretically fully explicit and embodied in a systematic theory of distributive justice.

More generally, my skeptical view is that allocation can be rational and can be just, but there is an intuitive gap between general theory and detailed practical decisions that can never be closed by formal theory. This is a fact about human activity that cannot be ignored in our thinking about rationality and justice, whether in general or as applied to particular problems of concern, such as those of allocation of resources to scientific research. For some general supporting arguments for this view, see Suppes (2003).

References

Arrow, K. *Social Choice and Individual Values.* New York: Wiley, 1951.

Bruns, H. "Über die Integrale des Vielkörper-Problems." *Leipzig Berichte der Königlichen Sächsischen Gesellschaft der Wissenschaften: Mathematische Klasse,* pp. 1-39, 55-82, 1887.

Kant, I. *Critique of Pure Reason.* New York: Cambridge University Press. First published in 1781. Translated by P. Guyer and A.W. Wood, 1997.

Murphy, A. H., and R. L. Winkler. "Probability Forecasting in Meteorology." *Journal of the American Statistical Association.* 79: 489-500, 1984.

Smale, S. "The Fundamental Theorem of Algebra and Complexity Theory." *Bulletin of the American Mathematical Society (N.S.).* 4: 1-36, 1981.

Suppes, P. "Rational Allocation of Resources to Scientific Research." In L.J. Cohen, J. Los, H. Pheiffer, and K.-P. Podewski (eds.), *Logic Methodology and Philosophy of Science, IV,* 773-789. Amsterdam: North-Holland, 1982.

Suppes, P. "Philosophy of Science and Public Policy." In P.D. Asquith and P. Kitcher (eds.), *PSA 1984, Vol. 2,* pp. 3-13. East Lansing, MI: Philosophy of Science Association, 1984.

Suppes, P. "Rationality, Habits and Freedom." In N. Dimitri, M. Basili and I. Gilboa (eds.), *Cognitive Processes and Economic Behavior,* 137-167. New York: Routledge, 2003.

Selected Bibliography

Authored Books

Decision Making: An Experimental Approach. (with D. Davidson and S. Siegal) Stanford, CA: Stanford University Press, 1957; reprinted Midway Reprint, 1977, Chicago: University of Chicago Press.

Introduction to Logic. New York: Van Nostrand, 1957; reprinted New York: Dover, 1999.

Markov Learning Models for Multiperson Interactions. (with R. C. Atkinson) Stanford: Stanford University Press, 1960.

Axiomatic Set Theory. New York: Van Nostrand, 1960; slightly Revised Edition, New York: Dover, 1972.

First Course in Mathematical Logic. (with S. Hill) New York: Blaisdell, 1964; reprinted, New York: Dover, 2002.

Experiments in Second-language Learning. (with E. Crothers) New York: Academic Press, 1967.

Computer-Assisted Instruction: Stanford's 1965-66 Arithmetic Program. (with M. Jerman and D. Brian) New York: Academic Press, 1968.

Studies in the Methodology and Foundations of Science: Selected Papers from 1951-1969. Dordrecht: Reidel, 1969.

A Probabilistic Theory of Causality. Acta Philosophica Fennica, 24. Amsterdam: North-Holland, 1970.

Foundations of Measurement, Vol. I: Additive and Polynomial Representations. (with D. Krantz, R. D. Luce, and A. Tversky) New York: Academic Press, 1971; reprinted, New York: Dover, 2007.

Computer-Assisted Instruction at Stanford, 1966-68: Data, Models, and Evaluation of the Arithmetic Programs. (with M. Morningstar) New York: Academic Press, 1972.

The Radio Mathematics Project: Nicaragua, 1974-1975. (with B. Searle and J. Friend) Stanford, CA: Stanford University, Institute for Mathematical Studies in the Social Sciences, 1976.

Logique du Probable. Paris: Flammarion, 1981.

Probabilistic Metaphysics. Oxford, England: Blackwell, 1984.

Estudios de Filosofia y Metodologia de la Ciencia. Alianza Universidad, S.A., Madrid, 1988.

Foundations of Measurement, Vol. II: Geometrical, Threshold, and Probabilistic Representations. (with D. H. Krantz, R. D. Luce, and A. Tversky) New York: Academic Press, 1989; reprinted, New York: Dover, 2007.

Foundations of Measurement, Vol. III: Representation, Axiomatization, and Invariance. (with R. D. Luce, D. H. Krantz, and A. Tversky) New York: Academic Press, 1990; reprinted, New York: Dover, 2007.

Language for Humans and Robots. Oxford, England: Blackwell, 1991.

Models and Methods in the Philosophy of Science: Selected Essays. Dordrecht: Kluwer Academic Publishers, 1993.

Language and Learning for Robots. (with C. Crangle) Stanford, CA: CSLI Publications, 1994.

Foundations of Probability with Applications: Selected Papers, 1974-1995. (with M. Zanotti) Cambridge: Cambridge University Press, 1996.

Representation and Invariance of Scientific Structures. Stanford, CA: CSLI Publications, 2002.

Elementary and Middle School Textbooks

Geometry for Primary Grades: Book 1. (with N. Hawley) San Francisco: Holden-Day, 1960.

Geometry for Primary Grades: Book 2. (with N. Hawley) San Francisco: Holden-Day, 1960.

Sets and Numbers. (Teacher's Edition, Books K-6). New York: Random House, 1966.

Sets and Numbers. (Books K-2). New York: Random House. Revised Edition, 1968.

Sets and Numbers. (Books 3-6). New York: Random House. Revised Edition, 1969.

Individualized Mathematics, Drill Kits AA-DD. (with M. Jerman) New York: Random House, 1969.

Sets, Numbers, and Systems: Books 1 and 2. (with B. Meserve and P. Sears) New York: Random House, 1969.

Sets, Numbers, and Systems, Teacher's Edition, Book 1. (with B. Meserve and P. Sears) New York: Random House, 1969.

Sets, Numbers, and Systems, Teacher's Edition, Book 2. (with B. Meserve and P. Sears) New York: Random House, 1970.

Edited Books

The Axiomatic Method with Special Reference to Geometry and Physics. (Proceedings of an international symposium held at the University of California, Berkeley, December 16, 1957 – January 4, 1958.) Co-edited with L. Henkin and A. Tarski. Amsterdam: North Holland, 1959.

Mathematical Methods in the Social Sciences. Co-edited with K. J. Arrow and S. Karlin. Stanford: Stanford University Press, 1959.

Logic, Methodology, and Philosophy of Science. (Proceedings of the 1960 International Congress.) Co-edited with E. Nagel and A. Tarski. Stanford: Stanford University Press, 1962.

Mathematical Methods in Small Group Processes. Co-edited with J. H. Criswell and H. Solomon. Stanford, CA: Stanford University Press, 1962.

Aspects of Inductive Logic. Co-edited with J. Hintikka. Amsterdam: North-Holland, 1966.

Philosophy, Science, and Method: Essays in Honor of Ernest Nagel. Co-edited with S. Morgenbesser and M. White. New York: St. Martin's Press, 1969.

Research for Tomorrow's Schools: Disciplined Inquiry for Education. Co-edited with L. Cronbach. New York: Macmillan, 1969.

Information and Inference. Co-edited with J. Hintikka. Dordrecht: Reidel, 1970.

Space, Time, and Geometry. Dordrecht: Reidel, 1973.

Approaches to Natural Language. Co-edited with K. J. J. Hintikka, and J. M. E. Moravcsik. Dordrecht: Reidel, 1973.

Logic, Methodology, and Philosophy of Science IV. (Proceedings of the Fourth International Congress for Logic, Methodology, and Philosophy of Science, Bucharest, 1971.) Co-edited with L. Henkin, G. C. Moisil, and A. Joja. Amesterdam: North-Holland, 1973.

Contemporary Developments in Mathematical Psychology, Vol. 1: Learning, Memory and Thinking. Co-edited with D. H. Krantz, R. C. Atkinson, and R. D. Luce. San Francisco: Freeman, 1974.

Logic and Probability in Quantum Mechanics. Dordrecht: Reidel, 1976.

The Radio Mathematics Project: Nicaragua 1976-1977. Co-editor with B. Searle and J. Friend. Stanford, CA: Institute for Mathematical Studies in the Social Sciences, Stanford University, 1978.

Impact of Research on Education: Some Case Studies. Washington, DC: National Academy of Education, 1978.

Studies in the Foundations of Quantum Mechanics. East Lansing, MI: Philosophy of Science Association, 1980.

University-Level Computer-Assisted Instruction at Stanford, 1968-1980. Stanford, CA: Stanford University, Institute for Mathematical Studies in the Social Sciences, 1981.

Ancient & Medieval Traditions in the Exact Sciences: Essays in Memory of Wilbur Knorr. (CSLI Lecture Notes). Co-edited with H. Mendell, and J. M. Moravcsik. Stanford, CA: CSLI Publications, 2000.

Stochastic Causality. Co-edited with M. C. Galavotti, and D. Costantini. Stanford, CA: CSLI Publications, 2001.

Reasoning, Rationality, and Probability. Co-edited with M. C. Galavotti, and R. Scazzieri. Stanford, CA: CSLI Publications, 2008.

Selected Articles

"The Role of Subjective Probability and Utility in Decision-Making." *Proceedings of the Third Berkeley Symposium on Mathematical Statistics and Probability, 1954-1955.* 5: 61-73, 1956.

"Axioms for Relativistic Kinematics with or without Parity." In. L. Henkin, P. Suppes, and A. Tarski (eds.), *The Axiomatic Method with Special Reference to Geometry and Physics.* Proceedings of an international symposium held at the University of California, Berkeley, December 16, 1957 - January 4, 1958, pp. 291-307. Amsterdam: North-Holland, 1959.

"A Comparison of the Meaning and Uses of Models in Mathematics and the Empirical Sciences." *Synthese.* 12: 287-301, 1960.

"Probability Concepts in Quantum Mechanics." *Philosophy of Science.* 28: 378-389, 1961.

"Stimulus-Response Theory of Finite Automata." *Journal of Mathematical Psychology.* 6: 327-355, 1969.

"Semantics of Context-Free Fragments of Natural Languages." In K. J. J. Hintikka, J. M. E. Moravcsik, and P. Suppes (eds.), *Approaches to Natural Language*, 370-394. Dordrecht: Reidel, 1973.

"Is Visual Space Euclidean?" *Synthese.* 35: 397-421, 1977.

"Propensity Representations of Probability." *Erkenntnis.* 26: 335-358, 1987.

"The Transcendental Character of Determinism." In P. A. French, T. E. Uehling, and H. K. Wettstein (eds.), *Midwest Studies in Philosophy, Vol. XVIII*, 242-257. Notre Dame, IN: University of Notre Dame Press, 1993.

"The Nature and Measurement of Freedom." *Social Choice and Welfare.* 13: 183-200, 1996.

"Invariance Between Subjects of Brain-Wave Representations of Language." (with B. Han, J. Epelboim, and Z.-L. Lu), *Proceedings of the National Academy of Sciences, USA.* 96: 12953-12958, 1999.

All of Patrick Suppes' articles can be found online at: http://suppes-corpus.stanford.edu

18
Nancy Tuana

DuPont/Class of 1949 Professor of Philosophy, Science, Technology, and Society, and Women's Studies

Director, Rock Ethics Institute, Penn State University, USA

Question 1—The Question of Origins

The power and potential of a Quinean approach triggered my early passion for philosophy of science. I took seriously Quine's mandate in "Naturalizing Epistemology" that we abandon both the quest for rational reconstruction in the sciences or in any other realm of knowledge production, as well as cease our efforts to prove that our beliefs are derived from certain foundation, and turn rather to a study of how we in fact form our beliefs and advance knowledge practices.

My graduate education occurred at a heady time. I began my studies in the early 70s when Kuhn's *Structure of Scientific Revolutions* was being read with both interest and passion by those of us studying science and when women's studies courses were just beginning to be offered on university campuses. Both Kuhn and the methods emerging from the new field of women's studies influenced my interpretation of Quine's charge that we study knowledge production. I audited as many courses as I took for credit, dividing my time between the physics courses that would inform my dissertation project and the women's studies courses that would inform my passions.

Financial pressures led me to take on full time employment as soon as I had completed my course work, moving across the country between one-year gigs while I wrote my dissertation. Each of these jobs had a profound influence on my research trajectory. In my first position I was asked to develop a women's studies program for a small community college in Northern California, which provided me with a breath of understanding of this newly developing field and a passion for feminist research that became the axis of my work. In my second position I had the great good fortune to work with John McDermott and Larry Hickman. Still auditing classes even while a full time faculty member, their classes and long hours of conversation sparked my love of the work of Dewey. Now influenced by the work of Dewey and the insights of Kuhn, and committed to feminist methods, I read Quine's proposal that we naturalize epistemology not as a clarion call to neuropsychology, which, arguably, may have been the more accurate interpretation of Quine's intent, but rather as a very early injunction to "situate" knowledge practices. Yet another job, now my first tenure track position, provided me the opportunity to work with George McClure who introduced me to the work of Whitehead as well as the opportunity to meet and learn from the biologist Richard Lewontin, R..

A Quinean/Deweyan/Whiteheadean inspired philosophy of science led me to take seriously a series of tenets that created the groundwork for my efforts in contributing to a feminist philosophy of science. Quine's coherentist theory of truth and his related tenet in "Two Dogmas" that truth emerges out of an inextricable interaction of language and extra-linguistic fact, opened up the range of relevant questions to be asked in examining scientific theories and beliefs. While not denying the impact of experience, the fine line ontologically delimiting language and materiality dissolves. Dewey's emphasis on the interaction between organism/environment/culture in which each component is co-constituted through their interactions and Whitehead's process metaphysic dissolved other boundaries. Quine and Dewey's reminders of the robust exchanges between science, common sense, and metaphysics set the stage for a rich appreciation of the social and political factors that are part of knowledge practices, as well as the rejection of a uniquely true or best supported practice of science. On such a conception, science is always a cultural practice, shaped in part by the ontologies and folk wisdoms of its contexts. In other words, knowledge is always local knowledge.

This was the philosophical background against which I read feminist work in the area of science. Carolyn Merchant's *Death of Nature* (1980) exposed the role of gendered beliefs in the construction of modern science and enabled me to begin to see the impact of sexism and androcentrism upon scientific theorizing and practice. Harding's and Hintikka's early volume, *Discovering Reality* (1983) provided a model for how feminist methods could inform and transform philosophy of science. Indeed, it was through a feminist theoretical lens and attention to the sex/gender system in science that a final divide, that between fact and value, began to blur. In this way I began to understand far more robustly than dreamed of in the philosophies of Quine, or arguably even Dewey, how richly knowledge is in fact situated.

It was just at this time in the development of my work in feminist philosophy of science that Peg Simons who was editing *Hypatia* invited me to guest edit a special issue of *Hypatia* on the topic of feminism and science. The call for papers I issued in 1985 resulted in such a robust response that we ended up with two special issues on the topic that were quickly turned into the top selling *Hypatia* book, *Feminism and Science,* which is still being regularly used in philosophy and women's studies classes. This amazing opportunity enabled me to create strong ties to the top theorists in this emerging field of study, beginning dialogues that continue to this day.

While it was Quine, and later Dewey and Whitehead, who originally drew me to philosophical issues regarding science, it was feminist science scholars who sustained my passion. Our attention to the relationship between power and knowledge in the sciences and our careful tracings of the role of social beliefs and values upon the practice of science gave rise to a wealth of new questions and opened up philosophy of science to new approaches and contexts of inquiry.

Question 3—Contributions

The range of questions raised and new methodologies developed by feminist philosophers of science and science studies theorists since we began our examinations of science in the early 80s is remarkable. Feminist philosophy of science has, from its outset, been informed by and developed in concert with the work of feminist scientists such as Evelyn Fox Keller, Anne Fausto-Sterling, Donna Haraway, and Ruth Hubbard, as well as feminist historians and sociologists of science. The dissolution of theoretical divides mentioned above—factual/linguistic, science/metaphysics/common sense, fact/value—also eradicated any firm divide between internal and external factors in the development of scientific knowledge and opened feminist philosophy of science to the social dimensions of scientific knowledge.

Our early work often began from careful examination of the complex ways in which scientific practice has been, and continues to be, informed by androcentric and sexist biases. From "man the hunter" theories of evolution to justifications of women's inferiority based on sex differences in brain lateralization, science often incorporated social biases about sex/gender and frequently served as the basis for justifying differential treatment of women and men and the gender roles accepted at that historical time period.

Realizing the extent to which such biases informed scientific theorizing and practice, feminist philosophers shifted from the more traditional questions, "How do we (or should we) develop scientific knowledge," to inquiries that reflect the ways in which science is situated within society/ies and the relationships between power and knowledge. In this our queries were both critical and constructive. Our questions became: "Why do we know what we know?" "Why don't we know what we don't know?" "Who benefits or is disadvantaged from knowing what we know?" "Who benefits or is disadvantaged from what we don't know?" "Why is science practiced in the way that it is and who is advantaged or disadvantaged by this approach?" "How might the practice of science be different?"

These inquiries led many of us to question the extent to which traditional models of rationality and/or the scientific method have privileged traits viewed as masculine and excluded and/or denigrated those perceived to be feminine. And once we became aware of the role of gender bias in science the tangle of other forms of bias, assumptions concerning the inferiority of certain races or the deviancy of certain forms of sexuality, for example, also became evident. A central research project of feminist philosophers of science was to examine whether or not correcting for bias, whether due to gender or to race or to some other difference, would require a transformation of scientific methods. Our investigations also included: examining the impact of gendered language and concepts, as well as beliefs about gender roles upon scientific categories and practices (e.g., Potter's argument that gender politics shaped the articulation and confirmation of Boyle's gas laws); opening scientific methods to neglected questions (e.g., research on menopause); taking account of excluded groups and/or knowledge practices (e.g., the study of so-called indigenous knowledge practices[1]). Our goal was not only to support a more inclusive science, but to contribute to the development of *better* science. For

---

[1] As Sandra Harding has argued, all knowledge practices, including Western science are indigenous knowledge practices, shaped by the interests and contexts of their development.

feminists, as for other liberatory theorists, better science, indeed better knowledge practices, are those that contribute to a more just society.

My contributions to the development of feminist philosophy of science cluster in four areas:

1. The role of values in science—both in terms of tracing the impact of gender politics on the practice of science as well as examining which values would support liberatory practices of science.

2. Questioning the sex/gender distinction in science and in feminist theorizing.

3. Contributing to a more adequate metaphysic.

4. Developing approaches to philosophy of science that are socially relevant.

1. The Role of Values in Science. In *The Less Noble Sex* I examine the ways in which scientists' metaphysical inheritance from religious and philosophical systems of belief influenced their empirical investigations of woman's nature and, in turn, reinforced and perpetuated those same systems of beliefs. But I also argue that this influence cannot be easily extricated because the same theories associate certain traits—reason, objectivity, and morality—with males, or, to be more accurate, those males who are seen as most fully evolved. Here as in many other instances the intersectionality between gender and race is very clear.

In *The Less Noble Sex* I examine a cluster of mutually supporting tenets that remained relatively constant in Western belief systems: that woman is by nature less perfect than man, in particular that she possesses inferior rational capacities and a defective moral sense and, as a result, requires social control both for her own good and for the good of society. The belief that woman is less perfect is also often associated with the tenet that man is the primary creative force, in all meanings of that phrase. I argue that the conception of woman as less than man, less perfect, less evolved, less divine, less rational, less moral, less healthy, less creative, is more than simple bias, easily amenable to revision. It is rather part of our inherited metaphysic that permeates our social, cultural, and political institutions. Furthermore, the persistent rationalization offered to justify and account for woman's inferiority, namely, woman's role in reproduction as contributing to her inferior development, remains constant despite variations on particulars.

Rather than arguing that science that is shaped by values is thereby bad science, my analysis embraced the work of theorists like Sandra Harding, Helen Longino, and Lynn Hankinson Nelson who argued that values are an integral part of knowing and cannot be purged from knowledge practices, including science. Good science, like any adequate knowledge practice, must rather systematically identify and examine the adequacy of the values it incorporates and supports.

In "Re-Valuing Science" I argue in concert with the standpoint perspective articulated by Sandra Harding that the experiences of women, particularly those who are scientists, provide resources for feminist rejections of the ideal of value-neutrality in science as well as raising questions about the epistemic commitments of traditional views of scientific practice. In particular, I take issue with the epistemic model of scientists as detached, disinterested individuals and argue instead for a dynamic model of engaged, committed individuals-in-communities. I argue that scientific methods, beliefs, and practices emerge in concert with the beliefs and values—metaphysical as well as aesthetic and moral—of the larger communities of which scientists are a part.

Most recently I have used the concept of epistemologies of ignorance, first coined by Charles Mills in the context of racism, to provide a helpful lens for analyzing the role of values in science. In two recent papers, "The Speculum of Ignorance" and "Coming to Understand," I argue that a more robust understanding of the different types of ignorance enables us to better understand instances of not knowing that are actively produced and the values that underlie the cultivation of such forms of ignorance.

A commitment to what Helen Longino labeled "changing the subject" whereby we make a commitment to fully articulating *situated knowers* leads also to an appreciation of the role of affect and of embodiment in the knowledge process and argues against conceptions of rationality, in the sciences or other realms, that are divorced from desires and interests. In this, feminist attention to the inextricable interconnections between individuals led many feminist scholars to alternative models of knowledge, such as Lorraine Code's account of knowledge modeled after "knowing other people," which she argues is better suited to instances where knowledge requires constant learning, is open to interpretation at various levels, admits of degree, and is not primarily propositional. In this I and many other feminist theorists argued for accounts of scientific knowledge that recognize the importance of empathy and imagination in the epistemic process.

As we situate knowers, it quickly becomes clear that it is not only the passions which must be included within any adequate account of scientific knowledge, but also embodiment. And once we recognize embodiment, we must also attend to the ways in which our material differences are epistemically significant.

2. Questioning the Sex/Gender Distinction. One thread that has remained constant since I began working on topics in feminist philosophy of science was a fundamental rejection of the sex/gender distinction. In one of my earliest publications in feminist philosophy of science published in the *Hypatia* special issue of *Women's Studies International Forum* in 1983, I argued that the sex/gender distinction, while arguably instrumental in launching feminist critiques of certain forms of discrimination against women, was highly pernicious. I document that it was not only a key component of scientific accounts of women's and non-white men's inherent inferiority, but is based on an inadequate ontology. In this and my later work, I argue that the nature/nurture distinction which grounds the sex/gender distinction is a key axis for theories of biological determinism that are both discriminatory and empirically inadequate.

While the links between biological determinism and discrimination are less surprising, I also argue that a more unanticipated consequence of the feminist embrace of the sex/gender distinction has been a problematic neglect of embodiment in feminist theorizing. I argue that many feminists unwittingly repeat a fundamental tenet of biological determinism in assuming that those features arising from sex, unlike those arising from gender, are not malleable. Having made such a move, feminists are then compelled to minimize biology—arguing that there are no significant biological differences between women and men (or other others).

Embracing the sex/gender distinction also leaves feminist theorizing open to the critique that biology is more significant than we claim it to be, and, indeed, inevitable; a critique often repeated in science as well as the popular media. I argue that feminists have been epistemically irresponsible in leaving in place a fixed, essential, material basis for human nature, a basis that renders biological determinism meaningful. Influenced by the work of Sarah Lucia Hoagland, I argue that our efforts must not be aimed at *disproving* biological determinism, but rather at rendering it *non-*

sense.

3. *A More Adequate Metaphysic.* In developing my critique of the sex/gender distinction and working to understand both how the body is socially constituted and how its materiality in turn informs the parameters of its configurations I come to appreciate the inadequacy of a physical object ontology. Influenced by Haraway, Dewey, Whitehead, and Lewontin, R., I argue for a new metaphysic, what I call *interactionism*, in which an organism and its environment are not separate and dichotomous processes, but always dynamically interrelated in phenomena, where phenomena are simultaneously material-semiotic. An interactionist account stresses not fixed entities but rather "dynamic interactions at multiple levels, from organisms with environments, to bodies and cultures, to those of DNA with the molecules and ions in a cell, where that cell is itself in interaction with its environment" (1996. 65).

Given an interactionist ontology, biological determinism and its mirror opposite, versions of social constructivism that posit bodies as simply cultural constructs, are nonsense. Interactionism undermines the either/or of biological determinism/social constructivism by rendering meaningless the poles of this dichotomy. Bodies are theorized (and lived) as material-semiotic phenomena.

The processes denoted by terms such as sex/gender or any related pairing such as nature/nurture or organism/environment are not separate and dichotomous processes, but always dynamically interrelated in phenomena. I stress that it does not follow from this that we cannot make distinctions, even distinctions between sex and gender, but that we must be aware of the situatedness of any such distinction. Each such distinction emerges out of the complex interactions between bodies and meanings, values and embodiments, organisms and environments, where the very processes connoted by each term are themselves emergent. Given this, I argue that we must become cognizant of the distinctions we habituate and construct, and advocate more epistemically responsible distinctions. We here see the truth of what Lorraine Code has so persuasively argued, namely that we cannot separate epistemic analysis from ethical analysis. To know well, we must be responsive to the differences articulating themselves in our experiences and practices, along with being attentive to how the distinctions we embrace, in part, construct our experiences, as well as how these distinctions are enacted in social practices, how they enable as well as limit possibilities and for whom, what they conceal as well as what they reveal, and so on.

While it was feminist theorizing that led me to the need to embrace interactionism, it became a lens for better understanding and more responsibly theorizing many phenomena, from climate change to plastic to the city of New Orleans. In my work on the impact of Hurricane Katrina, I argue for the importance of embracing an ontology that rematerializes the social and takes seriously the agency of the natural. I argue that interactionism draws our attention to relatively neglected or poorly understood phenomena, namely, knowledge of the interaction between things and people, between feats of engineering and social structures, between experiences and bodies. I argue that what is often inadequately understood is the porosity of the complex interrelations from which phenomena emerge.

The phenomena of Katrina, the levee system in New Orleans, and polyvinyl chlorides all argue for the inadequacy of contemporary debates between versions of realism and social constructivism. In "Witnessing Katrina" I argue that Katrina is a natural phenomenon that is what it is in part because of human social structures and practices. It and the levee system in New Orleans, the Mississippi, polyvinyl chlorides, and cancer, to name just a few phenomena, are emblematic of the porosity between humans and our environment, between social practices and natural phenomena. And through this lens I call for careful attention to the complexities of material agency and the viscous porosity of flesh—my flesh and the flesh of the world. And again I argue that nature/culture is a problematic ontology: not just for the human world, but for what is, as well as what might yet be.

4. The Importance of a Philosophy of Science that is Socially Relevant. Feminist philosophy is at heart a liberatory philosophy. Its ultimate goal is to contribute to the formation of a more just society. Feminist philosophy of science and feminist science studies have focused on uncovering sources of ignorance and error that have contributed to practices of oppression as well as urging efforts to use science to enrich flourishing of not only women, but of all humans in all countries as well as the flourishing of the environment. Feminist philosophy of science contributes to this goal by carefully examining and evaluating the linkage between power/knowledge and the role of politics in science to carefully examine how science can best contribute to such flourishing.

Originally influenced by Quine's and Dewey's philosophies of "rubbing out boundaries," my most recent work is devoted to rethinking and undoing the divisions between philosophy of science and ethics, to building bridges between philosophy of science and environmental policy, and bringing philosophy of science to policy. Over the last four years through a focal initiative of the Penn State Rock Ethics Institute, I have devoted my efforts to exploring ways to make an impact both nationally and globally on climate change policy. Working with climate scientists at Penn State and collaborating with an international group of climate scholars and negotiators, I am working to bring the methods of philosophy of science and ethics to bear on how climate science can more responsibly inform climate policies. These collaborations involve working with Penn State climate scientists to develop cutting-edge policy-relevant research as well as the development of the Collaborative Program on the Ethical Dimensions of Climate Change (EDCC) designed to bring together an international group of philosophers, climate scientists, and policy-makers in order to insure that local, national, and global climate policies are epistemically responsible. The EDCC has been participating in the annual United Nations Framework Convention on Climate Change and we are currently working to have an impact on the next report of the Intergovernmental Panel on Climate Change. We have also launched ClimateEthics.org, which was recently named by *Time/CNN* as one of the top 15 Green Websites in the world. ClimateEthics.org was created to generate ongoing ethical analyses of emerging considerations in climate policy development and the science that supports these policies. My work also focuses on the gender dimensions of climate change vulnerabilities and impacts.

The Vital Question

The remaining three questions—what are the most interesting, important, or pressing problems in philosophy of science/what is the relation between philosophy of science and efforts for social justice/where do you see philosophy of science heading—are in my view so closely interlocked that they cannot be separated. As I think about the role of the academy and of the work that we do as philosophers of science, there are no more interesting or pressing problems we face as philosophers than the question of how our work can help shape science and policy that contributes to flourishing, of humans locally and globally, as well as the flourishing of the environment. To address human flourishing, philosophy of science must be attentive to the ways in which our research can benefit efforts for social justice. The flourishing of the nonhuman environment is in turn linked to human flourishing, for impoverished ecosystems are all too often sources of harms to humans. Here the division between philosophy of science and environmental philosophy must be carefully rethought.

Consider human flourishing. Efforts for social justice must begin by ensuring that basic rights to adequate food, clean water, health, and shelter are satisfied. While these are only the minimal foundation needed for other rights such as political security, education, individual, civil and political liberty, and the like, they are currently out of the reach of hundreds of millions of people. Science and Technology are a key component of effective efforts to ensure basic rights for all peoples. Philosophy of science must thus focus attention not only on how to ensure good science, but on science that is a resource for a more just society. In this, we can see the goals of an adequate philosophy of science as mirroring the NSF two criteria for judging research proposals. The first criterion focuses on the intellectual merit of the proposed activity with emphasis on the technical feasibility and creativity of the project. The second criterion emphases the broader impacts of the proposed activity including the potential benefits to society. Our work as philosophers of science must strive to contribute to *both* of these charges, to an enlarged view of philosophy of science. While feminist scholarship as explained above has greatly contributed to this enlarged view of philosophy of science, this is not a new vision of the philosophy of science. Indeed a plausible argument could be made that it is rather a contemporary return to the aims of philosophy of science as it emerged out of the Vienna Circle.[2]

The urgency of shaping philosophy of science to provide resources for addressing social and environmental issues is illustrated by the extent of the global problem that we face at the beginning of the twenty-first century. To provide just one simple snapshot of these issues, consider the recent list of the ten most pressing problems facing the world today as identified by the 2008 *Copenhagen Consensus*:

- Air Pollution
- Diseases
- global warming
- Sanitation and water
- Terrorism
- Conflicts
- Education
- Malnutrition and hunger
- Subsidies and trade barriers
- Women and development

---

[2] See, for example Cartwright et al, 1996.

While not all may agree with every item on this list, there is nevertheless strong consensus on many of them: from the importance of access for all to adequate food and clean water, to reducing the number of deaths from easily controlled diseases. Just a few facts about poverty and malnutrition serve to underscore the nature and extent of the problem: Every five seconds a child dies because of malnutrition.[3] While malnutrition is often an underlying factor, millions of other children and adults die each year from diarrhea and respiratory infections such as pneumonia and bronchitis, deaths which would have been easily preventable through inexpensive medical treatments (such as antibiotics) and environmental interventions that would greatly improve the quality of water, air and sanitation.[4] According to the World Bank, over one billion people live in extreme poverty (defined as living on less than one dollar per day) and extreme poverty is an underlying cause of more than 8 million deaths each year.

One of the objectives of the UN Millennium Development Goals, agreed to by all the world's countries and all the world's leading development institutions, (http://www.un.org/millenniumgoals/) is to reduce by half the numbers of individuals and families living in extreme poverty by 2015. The Millennium Development Goals provide a blueprint for action for philosophies of science. While the solutions to these problems will require far more than scientific or technological progress, philosophy of science can partner with scientists, engineers, and policy makers to better identify research agendas in the sciences and engineering that will most effectively address solutions to these issues, examine how to best structure and support science that aims for social justice, identify factors that advance or impede such goals, and the like. Solutions to problems like these will not be simple and will require collaborative multidisciplinary and interdisciplinary scholarship. It is essential that philosophers of science become robust contributors to these important research projects.

Conclusion

Philosophy of science has been committed to interdisciplinary research. We have taken it as part of our charge that our background includes training in the sciences as well as knowledge of the history

---

[3] United Nations Food and Agriculture Organization, *State of Food Insecurity in the World* 2006

[4] World Health Organization, *the World Health Report* 2007.

of our focal science. And many of us have included as part of this charge, knowledge of the social dimensions of the practice of our chosen sciences. To this feminist and other liberatory philosophies have added the imperative that our work be linked to efforts for social justice and that we take seriously how our work can inform more epistemically responsible science policy. I can imagine no more exciting or important task for philosophy in the twenty-first century than to take this step and enlarge the scope of philosophy of science in this way.

References

Cartwright, N., J. Cat, L. Fleck, and H. Chang. *Otto Neurath: Philosophy Between Science and Politics.* New York, NY: Cambridge University Press, 1996.

Code, L. *Ecological Thinking: The Politics of Epistemic Location.* Oxford University Press, 2006.

Code, L. *What Can She Know?* Cornell University Press, 1991.

Haraway, D. *Primate Visions.* Routledge, 1989.

Harding, S. *The Science Question in Feminism.* Cornell University Press, 1986.

Harding, S. *Whose Science? Whose Knowledge?* Cornell University Press, 1991.

Harding, S., and M. Hintikka. *Discovering Reality: Feminist Perspectives on Epistemology, Metaphysics, Methodology, and Philosophy of Science.* D. Reidel, 1983.

Hoagland, S. L. "Resisting Rationality." In N. Tuana and S. Morgen (eds.), *Engendering Rationalities*, Indiana University Press, 2001.

Keller, E. F. *A Feeling for the Organism.* San Francisco: W.H. Freeman, 1983.

Keller, E. F. *Reflections on Gender and Science.* Yale University Press, 1985.

Kuhn, T. *The Structure of Scientific Revolutions.* Chicago: Chicago University Press, 1962.

Longino, H. *Science as Social Knowledge.* Princeton University Press, 1990.

Merchant, C. *Death of Nature: Women, Ecology, and the Scientific Revolution.* Harper and Row, 1980.

Mills, C. *The Racial Contract.* Cornell University Press, 1999.

Nelson, L. H. *Who Knows: From Quine to a Feminist Empiricism.* Temple University Press, 1990.

Potter, E. *Gender and Boyle's Law of Gases.* Indiana University Press, 2001.

Selected Bibliography

Selected Books

Feminism and Science. Editor. Bloomington: Indiana University Press, 1989.

The Less Noble Sex: Scientific, Religious and Philosophical Conceptions of Woman's Nature. Indiana University Press, 1993.

Selected Articles

"Re-fusing Nature/Nurture." *Hypatia*, published as a special issue of *Women's Studies International Forum.* 6(6): 45-56, 1983.

"Engendering Science: From the Perspective of the Humanities." *National Women's Studies Association Journal.* 5(1): 56-64, 1993.

"The Values of Science: Empiricism From a Feminist Perspective." *Synthese.* 104(3): 1-21, 1995.

"Fleshing Gender, Sexing the Body: Refiguring the Sex Gender Distinction." Spindel Conference Proceedings, *Southern Journal of Philosophy.* XXXV: 53-71, 1996.

"Re-Valuing Science." In L. H. Nelson and J. Nelson (eds.), *Feminism, Science, and the Philosophy of Science*, 17-38. Kluwer, 1996.

"Material Locations: An Interactionist Alternative to Realism/Social Constructivism." In N. Tuana and S. Morgen (eds.), *Engendering Rationalities*, Indiana University Press, 2001.

"Coming to Understand: Orgasm and the Epistemology of Ignorance." *Hypatia: A Journal of Feminist Philosophy.* 19(1): 194-232, 2004.

"The Importance of Expressly Integrating Ethical Analyses into Climate Change Policy Formation." (with D. Brown and J. Lemons) *Climate Policy.* 5: 1-4, 2006.

"The Speculum of Ignorance." *Hypatia: A Journal of Feminist Philosophy*, Special issue on Ethics and Epistemologies of Ignorance. 21(3): 1-19, 2006.

White Paper on the Ethical Dimensions of Climate Change, lead author with Don Brown, 2006.

"Human-Environment Interactions: A Plea for the Humanities." *Nature and Culture*. 2(2): 210-222, 2007.

"Interdisciplinary Studies in Science, Technology, and Society: New Directions: Science, Humanities, Policy." (with R. Frodeman, J. T. Klein, and C. Mitcham) *Technology in Society*. 29(2): 145-152, 2007.

"Viscous Porosity: Witnessing Katrina." In S. Hekman and S. Alaimo (eds.), *Maternal Feminisms*. Duke University Press, 2007.

About the Editor

Robert Rosenberger is an assistant professor of philosophy in the School of Public Policy at the Georgia Institute of Technology. He has written on a variety of topics in the philosophy of science and philosophy of technology, such as the conventions of scientific debate, and the habitual relationships people develop with everyday technologies such as cellular phones, desktop computers, and television. In a continuing series of articles, he investigates the roles played by imaging technologies in scientific practice, with case studies in neurobiology and the exploration of Mars. In addition, he an his colleagues in The Group for Logic and Formal Semantics study the philosophy of computer simulation in science, and have constructed game-theoretic models of social psychological theories of prejudice reduction.

Index

A-level chemistry, 16
actor-network theory (ANT), 110, 111
anti-realism, 51, 164, 190, 191

Barad, K., 186
Bayes, 36, 38, 181
    bayesian, vi, 36, 37, 39, 46, 47, 73, 161, 181
Beck, U., 95
bell inequalities, 186
biological species concept, 17

Carnap, R., 56, 57, 71, 73, 89, 90, 137, 138, 169, 183, 191
Cartwright, N., 76, 78, 81, 141, 144, 154, 186, 233
civic epistemologies, 120
climate change, v, 29, 59, 154, 193, 194, 196, 229, 230
    climate skeptics, 116
    global warming, 44, 59, 60, 167, 231
co-production, 119, 120, 123, 126
Code, L., 226, 228, 233
cognitive science, 11, 75, 190, 194
Cold War, 58, 94, 106, 110
conflict of norms, 204
conformation, 152
consciousness, vi, 100, 112, 127, 159, 166, 169

constitutive values, 149, 151
contextual values, 148, 149, 151, 152
creation science, 139
    intelligent design, iv, v, 6, 19, 44, 85, 116, 140
    creationism, 29, 44, 144
critical contextual empiricism, 149, 151, 152

Darwin, *see* evolution
Darwinism, *see* evolution
Dawkins, R., 8, 16, 24, 85, 86
demarcation, 63, 116, 117
Dennett, D., 14, 24, 80
Dewey, J., iv, 29, 141–144, 202, 222, 223, 228, 229
distributive justice, 208, 212, 213
divide et impera, 176
DNA, *see* gene
Douglas, H., 34, 154
Duhem, P., 87, 101, 150, 155

Einstein, A., 28, 30, 31, 50, 51, 54–57, 65, 66, 86, 200
embodiment, 103, 105, 226–228
epistemological cultures, 133
error-statistics, 37, 39
Evans, R., 4, 10, 11
Everett, 162, 166, 167, 169

Everettianism, 162, 166, 167
evolution, iv, vi, 6, 14, 16, 17, 20, 24, 25, 44, 45, 70, 78, 79, 81, 85, 103, 106, 135, 139, 140, 142, 145, 162, 169, 190, 194, 224
  darwin, 17, 24, 25, 46, 139, 140, 142, 145
  natural selection, 132, 139, 140
evolutionary psychology, 20
experimenter's regress, 7

femininity, v
feminism, 93, 97, 99, 101, 102, 135, 155, 173, 223, 233–235
  feminist, iv, 28, 78, 90–92, 96–98, 100–102, 106, 134, 136, 143, 148, 152, 156, 179, 186, 189, 221–227, 229, 231, 233–235
Feyerabend, P., 57, 90, 91, 100, 138, 149, 150, 152, 155, 158, 191
Feynman, R., 8, 56, 60
Foucault, M., 50, 62, 186, 198

gender, v, 8, 92, 97, 101, 106, 130–132, 135, 156, 173, 222–225, 227, 228, 230, 233, 234
gene, vi, 16, 18, 20, 21, 24, 26, 101, 132, 133, 135, 145
  genome (Human Genome Project), 21, 60, 79, 106, 141
  DNA, 17, 45, 75, 79, 106, 126, 127, 228

global warming, *see* climate change
Goodman, N., 117, 120, 125, 137

Hacking, I., 40, 41, 43, 46, 71, 75, 105, 118, 125, 141, 157, 158, 169, 186
Hanson, R., 79, 150, 155
Haraway, D., 106, 186, 223, 228, 233
Heidegger, M., 51, 185, 192, 198
Hempel, C., iv, 83, 84, 137, 138, 149, 155, 169, 191
hermeneutic, 105, 107, 108
Hesse, M., 51, 162
human sociobiology, 140
Hurricane Katrina, 229

inference to the best explanation, v, 183
intelligent design, *see* creation science
interactional expertise, 7
interactionism, 228, 229

Kant, I., 51, 65, 202, 205, 214
Knorr-Cetina, K., 186
Kuhn, T., iv, 2, 9, 22, 28, 34, 46, 50, 51, 54, 55, 57, 58, 65, 70, 77, 79, 80, 89, 91, 97, 98, 100, 137–140, 144, 149, 150, 152, 155, 158, 180, 188, 190, 191, 198, 221, 222, 233

Lakatos, I., 50, 51, 65, 180
Lamarck, 17

Index   241

Latour, B., 8, 66, 105, 111, 113, 186
Lewontin, R., 134, 222, 228
logical positivism, 56, 57, 89, 90
logical empiricist, iv, 83, 84, 138
Vienna Circle, 27, 28, 56–58, 83, 231

many worlds, vi
Marcuse, H., 49, 65
Marx, 171
marxism, 56
masculinity, v, 130
Mayo, D., 37–40, 46, 72
Merchant, C., 222, 234
metaphor, vi, 131, 140, 141
microbe, vi, 17, 18, 20, 22
Mills, C., 226, 234
modernity, 95, 97–100, 102, 171

Nagel, E., 137, 201, 202, 217
natural ontological attitude (NOA), 28, 191
natural selection, *see* evolution, *see* evolution
Nelkin, D., 110, 111, 125
Nersessian, N, 75, 79
no miracles, 164, 175

objectivity, iii, v, 28, 62, 63, 65, 100, 104, 115, 116, 119, 120, 125, 130, 137, 150, 155, 164, 172, 174, 179, 225
Oreskes, N., 59, 60, 65

paradigm, 2, 106
perspectivism, 77, 80, 81
pessimistic meta-induciton, 164
pessimistic induction, 176, 177, 183

philosophy of biology, vi, 15–18, 20, 22, 24–26, 70, 72, 78, 79, 138–140, 142, 143, 186
philosophy of experiment, 35, 46
physicalism, 165, 166
physics, iv, vi, 7, 11, 18, 19, 27, 29–34, 39, 43–45, 47–56, 58, 59, 63, 65, 69, 70, 72, 73, 83, 84, 86, 88, 105–107, 123, 129, 135, 136, 138, 143, 157, 160, 163, 165, 166, 169, 171, 174, 185, 186, 190, 200, 201, 203, 205–209, 211, 217, 218, 221
Poincaré, H., 65, 209
Popper, K., 1, 22, 77, 158, 169
post-empiricist, 187, 188
post-positivist, 95
postcolonial, v, 92, 93, 96, 97, 99, 102
premise-circularity, 176
psychoanalysis, vi, 57, 86–88
psychology, vi, 20, 56, 61, 86, 88, 117, 138, 140, 143, 157, 159, 172, 174, 194, 203, 207, 218, 219, 222
Putnam, H., 31, 50, 84, 107, 164, 165, 169, 175, 191

quantum entanglement, vi, 206
Quine, W. V. O., iv, 28, 49, 55, 65, 90, 91, 98, 101, 137, 144, 169, 190, 191, 221–223, 229, 234

race, 8, 26, 142, 145, 154, 224, 225
racism, 92, 99, 226
realism, iv, v, 18, 19, 28, 30–32, 50, 51, 58, 70, 76, 77, 81, 103, 107, 112, 145, 158, 163, 164, 171, 172, 174–177, 182, 183, 190–192, 198, 229, 234
Reichenbach, H., 49, 65, 71, 84, 87, 90, 138
relativism, 28, 30, 58, 104, 112, 127, 158, 190
rule-circularity, 176
Russell, B., 57, 126, 150, 157, 200

science studies, iv, v, 4, 8, 10, 11, 24, 45, 58–60, 76, 96, 104, 106, 189–191, 223, 229
Science and Technology Studies (STS), 6, 60, 76, 81, 92, 96, 97, 100, 102, 109–114, 118–121, 123, 126, 127, 231
science wars, 104, 113
scientific revolution, 2, 34, 46, 50, 65, 70, 80, 89, 100, 107, 137, 144, 155, 221, 233, 234
sex, v, 11, 17, 93, 154, 222–225, 227, 228, 234
Smale, S., 211, 214
Smith, D., 90, 98, 99, 101
social construction, 97, 112, 125, 153, 190, 191, 228, 229
constructivism, 210
social science, 60, 90
sociology, 1, 2, 9–11, 13, 31, 61, 75, 98, 99, 101, 110, 115, 118, 125, 127, 139, 188–190
sociology of scientific knowledge
SSK, see strong programme
Sokal, A., 58
Solomon, M., 154
standpoint, v, 90, 98, 100, 102, 113, 118, 179, 201, 203–205, 210, 213, 226
strong objectivity, 98
strong programme, 110
sociology of scientific knowledge (SSK), 110, 113

theory dependence, 164, 165, 170
three-body problem, 209
Toulmin, S., 28, 138
trading zone, 55, 56, 60, 61

value-free, 141, 148
value-neutrality, 93, 148, 226
van Fraassen, B. C., 76, 77, 80, 182, 183
Vienna Circle, see logical positivism
Vietnam, 49, 58, 90, 110

well-ordered science, 141, 142
Winch, P., 1, 2
Wittgenstein, L., iv, 1–3, 9, 28, 54, 180

Yeats, W. B., 124

www.ingramcontent.com/pod-product-compliance
Lightning Source LLC
Chambersburg PA
CBHW021807220426
43662CB00006B/211